HUMAN ERROR IN COMPUTER SYSTEMS

HUMAN ERROR IN COMPUTER SYSTEMS

Robert W. Bailey, PhD.

PRENTICE-HALL, INC., *Englewood Cliffs, NJ 07632*

Library of Congress Cataloging in Publication Data

Bailey, Robert W.
Human error in computer systems.

Bibliography: p.
Includes index.
1. System design. 2. Human engineering. I. Title.
QA76.9.S88B34 1983 003 82-21492
ISBN 0-13-445056-6

Editorial/production supervision and
interior design: *Shari Ingerman*
Cover design: *Ray Lundgren*
Manufacturing buyer: *Gordon Osbourne*

Printed in the United States of America

10 9 8 7 6 5 4 3 2 1

ISBN 0-13-445056-6

Prentice-Hall International, Inc., *London*
Prentice-Hall of Australia Pty. Limited, *Sydney*
Editora Prentice-Hall do Brasil, Ltda., *Rio de Janeiro*
Prentice-Hall Canada Inc., *Toronto*
Prentice-Hall of India Private Limited, *New Delhi*
Prentice-Hall of Japan, Inc., *Tokyo*
Prentice-Hall of Southeast Asia Pte. Ltd., *Singapore*
Whitehall Books Limited, *Wellington, New Zealand*

To my sons
Chad, Gregg (Skip), Robert, and James

CONTENTS

PREFACE

This book began as a talk. In the early 1970s, I was assisting about forty designers in the development of a large, complex computer system. One day we learned that another new system being designed in the same organization was having serious difficulties in its first trial. We heard that a plane load of designers was flown 3000 miles to the trial site "to straighten things out." Some stayed for days, others for weeks, and many for *months* trying to make the necessary fixes. One of their most serious problems, and the one that finally forced the system to grind to a halt, was *excessive errors.* Not only the errors made by people preparing daily input, but also a large volume of new errors made by people who were *correcting* the original errors.

We wanted to avoid the same type of catastrophe. I was asked to prepare material on the sources, causes, and control of human error. After several weeks of gathering and organizing information, I made a presentation to the design team. It was discussed at great length, and eventually lead to several changes in our design strategies. The presentation then was repeated at least fifty times in the next twelve months for other system design groups. The popularity of the talk suggested that there was a need by many designers to understand issues related to human error in computer systems.

I soon tired of repeating the talk, although discussions that accompanied each presentation almost always produced new and interesting information. The presentation finally was converted into a talk monograph, and several hundred copies have been distributed over the past few years. This book updates and expands on the material in the original presentation and talk monograph.

The book is intended for computer system designers, or those learning to be designers. *The purpose of the book is to briefly convey several considerations related to controlling human error in systems.* It presents an overview of issues having to do with preventing, detecting, and correcting errors in computer-based systems. For a more detailed discussion of ways to prevent errors in systems, you are referred to my book, *Human Performance Engineering: A Guide for System Designers,* published by Prentice-Hall, Inc.

HUMAN ERROR IN COMPUTER SYSTEMS

1

DESIGNING TO REDUCE ERRORS

INTRODUCTION

In the development of many computer-based systems, there is much pressure on designers to make a *series* of correct design decisions. In essence, we frequently assume a "no second chance" design decision philosophy. Each decision builds on the one before it until they are stacked so high that it becomes almost impossible to alter early decisions. In the past the approach of simply "putting something together" and then trying it out to see how well it works was acceptable. Problems, particularly excess errors, were dealt with by taking time to redesign the system. This was an accepted, though not a very efficient way to develop systems. Times have changed.

Developing a new system and sitting back to see what happens (i.e. how many errors occur) costs too much in both time and money. Thus, it becomes important to make good design decisions from the very beginning, with *each* design decision being made to reduce the number of errors made in the system. For this reason it is important to know from the very beginning what kinds of decisions could result in high error rates.

Most all errors in the operation of computer-based systems can be attributed to the human, and many of these human errors can and should be controlled. At least half of the errors made during *operation* of a new system can be related to faulty decisions made during the *design* of that system. In fact, we are at a point in systems development technology where high error rates are usually considered as a sign of poor system design. Thus, one prime objective in developing a new data system should be to reduce human errors by modifying (in advance) the factors that produce them. In other words, we should design systems that minimize the potential of people to make errors.

HUMAN vs. COMPUTER PERFORMANCE

Most computer-based systems require both people and a computer in order to function properly. As computer systems have become more and more common, much attention has been focused on the reliability of these two main components — computer and people. Recent technological advances have greatly improved the reliability of the computer. In fact, when compared with the human, the computer is generally considered to be much more reliable. This comparison is frequently drawn, however, without similar systematic efforts being made to improve human reliability.

In one sense it is much easier to obtain and ensure an acceptable level of computer performance than human performance. This is because a designer usually knows in advance what the computer can and cannot, will and will not, do. In fact, we frequently design, develop and debug programs on the exact same computer that will be used in the new operational system. Unfortunately, we don't have the same opportunity with the people who are also going to be included in the system. Even if we were fortunate enough to know some of the new system users, we can generally assume that they will soon be transferred, promoted or quit. Thus, the people in our system will continually change, whereas the computer will probably remain essentially the same until even a newer system is introduced.

Some system designers have developed computer systems assuming that it's users will perform much like the computer, i.e., user performance will always be near perfect. In addition, some of these system designers naively believe that as long as the computer works, the people associated with the computer will be wildly excited and enthused at the opportunity of working in the system. They assume further that people in the new system will make few (if any) errors because the users want the computer to function at a high level. If users do make occasional errors it is because they are simply being "careless". These designers will tell you that carelessness is stopped by telling people to stop being careless. That is, you stop people from making errors by telling them to "stop making errors". All of these assumptions are usually untrue.

ERRORS AND DESIGN DECISIONS

To help obtain the level of human performance (i.e., minimal number of errors) required for their new systems, other (more informed) designers try to have people *selected* who already possess the necessary skills and knowledge concerning the things that need to be done. They do this by providing detailed descriptions of the types of people they had in mind for users when the design decisions were being made. If not too many of these people exist, which is frequently the case, then they recognize that someone is going to have to *train* people to do the job well. Thus, the designer ensures that adequate training is available. If adequate human performance is not achieved by way of selection and/or training, the deficiencies may be overcome, at least partially, by providing adequate instructions, including good performance aids. In general, however, adequate selection, training, and instructions lead to an acceptable level of human performance, only if all other human performance aspects of the system are *well designed.*

Even with a well designed system, it is important to recognize that all people do not make the same proportion of errors. In fact, some users may make up to *ten times more errors* than others (Klemmer & Lockhead, 1964). Even though it is well understood that some people make many more errors than others, most investigations of human error have traditionally used an "error-rate" approach where the emphasis was on uncovering a supposedly fixed level of human fallibility associated with a variety of manual activities, e.g., keypunching, coding, and calculating. That is, typical error rates were considered to be at a common level for most people, but differed for different tasks. This error rate approach is based on the acceptance of a certain amount of human error as being inevitable. Because of this assumption much of the research in this area was oriented toward *predicting* the human's contribution to system degradation. Unfortunately, this directed attention away from the numerous factors which gave rise to these errors, and to the possibility that knowledgeable design decisions could actually *prevent* many of these errors from ever occurring in the first place.

ERROR CONTROL

An error in a computer system is generally very explicitly defined. In its simplest terms, it is a wrong piece of data. Effective control over these errors should include provision for *prevention* of errors, *detection* of those that occur, and efficient *correction* of those detected. Truly effective error control in systems means much more than simply having the computer trying to detect errors. Unfortunately, over the past few years the major emphasis on error control has been on the computer detection of errors, rather than on the prevention or correction.

Considerations for reducing errors in a computer system should go far beyond the identification of errors by the computer. In fact, the most economical approach for dealing with most potential errors is to do whatever is necessary to prevent them from ever occurring at all. Consider that in most cases it takes only a second or two to make an error, but possibly as long as 15 to 30 minutes or more to detect and correct it (assuming that it can be detected).

There is a big difference between "error control" and "error detection". Error control refers to an extensive and well thought-out approach for minimizing human error in a computer system. As mentioned earlier, it includes attempts to prevent errors, to detect those errors not prevented, and to efficiently correct all errors found. Obviously, this includes much, much more than simply adding a few validations to a computer program. To concentrate only on error detection will severely limit the extent of accuracy control in most systems.

It's probably fair to say that if a system is designed where errors are not controlled (in advance), then the system itself could eventually reach the point of being "out of control." There have been some computer systems that were out of control to the extent that people had actually quit using the data produced by the systems. There have been systems that contained so many errors that "users" had built their own private (underground) manual systems, which they used instead of computer outputs. In other systems, designers have had to literally "pull the plug" on the computer, so that errors that had been piling up for days, months or even years, could be corrected. It is difficult to know in advance how much error is "too much," but for every system such a critical point exists. It is similar to listening to the radio. If the proportion of advertisements (e.g. errors) to music (e.g. correct data) is too high, we turn to another (alternative) station.

Errors tend to some extent or another to make a system less usable. In the extreme case, as was already mentioned, if the error rate for critical data items reaches a certain upper limit, the system will become unusable and consequently unused. To a lesser extent, all human errors that find their way into the system degrade the efficiency and effectiveness of the system.

THE "DETECTION-ONLY" PHILOSOPHY

As we have noted before, accuracy control means prevention, detection, and correction, but in most computer-based systems, we have tended to emphasize detection. A considerable amount of money is spent in trying to detect errors, especially with computer-based detection programs. In addition, correction procedures are generally oriented toward correcting each error as an individual error with little thought being given to why the errors occurred in the first place.

If a system designer concentrates only on developing software to detect errors in the computer, and ignores the sources and causes of error, the system will *never* function at an optimal level. A designer's consideration of error must extend back to the point where errors originate. And if records are used from outside the boundaries of the new system, they too must be considered and possibly "scrubbed" before being accepted for entry into the new system.

Any good system design effort will place considerable emphasis on the *prevention* of errors. This prevention process can continue even after the system is operational if the designer provides a means for identifying and altering the source of errors.

Excessive errors are symptoms of a sick system. A user cannot treat symptoms and hope to permanently cure a severe error problem. In most systems far too much time and money is spent by users having to correct errors. These are errors that could have been prevented by a designer. It's almost as though we think we can detect and correct enough errors to ultimately "do away" with the problem. This simply is not the case. Only by altering the *causes* of errors, hopefully while the system is being designed and not while it is being operated, can we hope to permanently reduce errors from occurring.

The "detection-only" philosophy usually begins with the designer; but once the system is operational the philosophy can extend from the user's top management down to the actual users of a system. Top management should require high accuracy levels, and should continually follow-up with detailed reviews of system performance. However, computer systems all too often operate with much less accuracy (higher error rates) than is necessary because management does not insists on strict error controls.

Unfortunately, too many people have become used to accepting a relatively high level of errors. A philosopher said many years ago, "What damned error, but some sober brow will bless it and approve it with a text." In most systems, we simply have not taken a good look to see how much the acceptance of high error levels is costing us. In fact, we sometimes go to great lengths to justify as inevitable many of the errors that are made in a system. A good system designer will incorporate an error control policy that tolerates few errors. In fact, a good designer will make a series of informed design decisions with "zero errors" as an objective.

ERROR COSTS

Error control is not without cost, but few designers truly understand the problems related to comparing the costs of having excessive errors with the costs of reducing them. In fact, it is probably safe to say that many designers feel much more comfortable "fixing errors", than trying to prevent them. In some organizations, large error correction departments have existed for years and continue to flourish.

Before implementing error controls, system designers should evaluate their merits by asking such questions as:

(a) What types of errors can be expected?

(b) How frequently might each error type occur?

(c) What are the cost consequences if the errors are not detected?

(d) What is the cost of preventing the errors of critical data items?

(e) What is the cost of detecting these errors?

(f) What is the cost of correcting these errors?

(g) What is the probability that each error type can be prevented and/or detected?

(h) If an error is not detected during computer or manual processing, what is the probability that it will be detected at a later stage?

(i) What are the cost and customer relation consequences of late rather than early detection?

One of the best ways to measure the effectiveness of a system to economically deal with errors is to determine the *cost per error.* A good estimate of the cost for *correcting* errors, for example, can be made by simply dividing the loaded salaries of the people who correct errors by the number of errors corrected in a certain time frame. For example, if you had one person who spent about eight hours correcting errors each day, and he or she made $5.00 per hour, it would cost about $40.00 per day to correct errors (5 x 8 = 40). If this person corrected an average of ten errors per day, the cost per error would be about $4.00 (40 ÷ 10 = 4). Certain errors may be corrected in five or ten minutes, while others will take 30 minutes to an hour or more.

The cost of correcting errors tends to increase in proportion to the time that elapses from when the error is made until it is detected. For example, it would probably take about ten seconds to correct an error that was noticed as a person was typing or keying, making the cost to correct that error only a few cents. If the error was detected as a result of proofreading, much more time would be involved. A proofreader would have to review the entire record in order to detect many errors, no errors, or just one error. With proofreading the error correction is already costing more, even though the original record is still readily available. This still makes it relatively convenient to find out what the correct answer should be. However, if errors are not detected until after the original record is filed or destroyed, the cost of making a correction jumps sharply. In addition, if an error turns out to be detected by a user or customer there could be the added "cost" of an irritated customer.

The cost of errors can be substantial. One study from a large communications company indicated that as many as 50 million human errors were made each year that adversely affected service to customers. They estimated the annual cost of correcting these errors to be over $100 million, or about $2.00 per error. Another study found that correcting *each error* was costing about $7.50. Still another study reported that correcting errors was costing about $8.00 each. Keep in mind that these studies considered only the cost of *correcting* errors that had been detected. The costs associated with having people or computers detect these errors is also substantial.

Some other errors have been costly. In one case, the lack of a ";" in the appropriate place in a computer program caused a United States satellite to go toward the Sun rather than Mars. The cost of this one simple error was placed at $80 million. In another instance, the lack of a parenthesis in a computer program caused the simultaneous destruction of several weather satellites (Chapanis, 1974).

A few years ago an international business ordered a special communications line for calling overseas. The service was properly installed and the customer was pleased. However, due to a human error, service was never billed. Five years later the error was detected, and the customer was presented with a $17,000 bill. The customer refused to pay and had the service disconnected.

An error in the telephone company Yellow Pages can be costly. For example, a justice of the peace in New Orleans sued the telephone company for $325,000 because his Yellow Pages listing improperly appeared under "junk dealers". The justice of the peace argued that the Yellow Pages error cost him $25,000, caused embarrassment, resulted in loss of reputation and made him the target of prank calls and insults.

In general, the cost of preventing most errors is significantly less than the cost of trying to detect and correct errors once they are made.

In one communications company, for example, the cost for performing equipment inventories and record reconciliations (i.e., the cost of matching the "real world" with the records in the office) was estimated to be at least $3.30 per telephone unit. In a typical telephone center with about 40,000 telephones, this amounted to $132,000 each time the records were reconciled. Unless significant changes were also made to the entire system at the same time the records were corrected, the file could easily require reconciliation every 3 to 5 years. Although substantial in and of itself, this figure did not take into account the cost of preventing, detecting and correcting errors on a day-to-day basis. All companies using computer-based records should periodically ensure that their database records match the "real world". And when they do, it is always a major expense. Much of this expense could be avoided if errors were never made in the first place.

Keep in mind that one of the most significant costs associated with errors occurs when data are converted from an existing manual system to a new computer-based data system. All errors made during this conversion process (and not detected and corrected) could float around in the new database "forever", plaguing all who try to use the system.

EXERCISES

The costs of errors can be estimated for many different situations. Once a "cost-per-error" is determined, then this cost can be traded-off against the cost of preventing the errors in the first place. Calculate the error costs in the following problems, and then compare the estimated costs for correcting the errors against the estimated cost for preventing them.

1. You manage a department of 12 employees who spend their day correcting computer-detected errors. Last year your staff corrected 112,000 errors. The total cost to your department for salaries and benefits was $627,200. What is the average cost per error?

2. The cost to redesign the system that produced the errors mentioned in Question 1 will be $1 million. If the system is redesigned, it is estimated that about half of the errors now being made will never occur. How much will the redesigned system save the company over a five year period? Remember to include the cost of redesigning the system.

3. Every year a special task force is created to find and reconcile errors in a 9,000 record database. This year you assign 6 people, who each make $200 per week. They finish the job in 12 weeks. These people must inspect each record, and compare what is on the record with actual conditions in the real world. At the end of the 12 weeks they have located and corrected 1058 errors. What is the average cost to reconcile each error?

4. You do a study and find that it takes an average of 15 *seconds* to correct certain types of keying error, but an average of 30 *minutes* to correct the same type of error two weeks later when it is detected in the database. The people keying data make $6 an hour, while the people correcting errors found in the database make $8 an hour. What is the average cost of correcting errors at each source?

2

MEASURING ERRORS

WHAT IS AN ERROR?

If the potential of the computer is to be realized, human error must be controlled. Errors made by people in computer-based systems are probably the single most significant deterrent to successful system operation. Most errors are made while preparing, entering, correcting or using data. An error may be the result of deviating from a prescribed procedure, making a wrong decision, striking the wrong key on a keyboard, or any of a number of other activities. In its simplest terms, a computer system error is a wrong piece of data.

In more general terms, an error may occur when a person fails to perform a task, when a task is performed incorrectly, when a task is performed out of sequence, or when a task is not completed within an allotted time period. We should be clear about how we are defining error. Human behavior may lead to performance errors, but the behavior itself is not the error. For our purposes, the consequence of the behavior, human performance in the form of incorrect or omitted data items, is the error.

Whether items of information are right or wrong is partially determined by their ability to fit in the space provided for the items in the computer's data base. The information *content* of the data may be right, for example, but the *format* may not fit data base requirements. On the other hand, the data may be totally wrong, but meet the format and content requirements of the computer and be accepted as "correct". In the latter case, the data items may be identified as being in error only when someone attempts to make use of them. To summarize, computer system errors may exist for one or both of the following reasons:

1. The data do not accurately represent "real world" conditions.
2. The data do not meet the content or format requirements of the system.

REPORTING ERRORS

How would you feel as a system designer, if all the errors from your systems were published every day like a baseball player's? Sometimes it seems that it would be useful for each system designer to be listed in the daily newspaper with yesterdays error rates for their systems listed next to their names. It wouldn't take too long to identify the good system designers. We could even carry it to the point like in baseball where the person with lowest error rate gets the "golden glove" award, the designer with the lowest error rate could get the "golden system" award. Unfortunately, in many systems reported error rates are so nebulous we simply could not compare one system with another even if we wanted to. It would be like each major and minor league baseball teams using a different formula and different criteria for calculating fielding errors.

How errors are reported in a computer system is important. Depending on your method, the error rate can either go up or down while the actual number of errors remains the same. There are at least three commonly used ways of reporting on errors. They can be reported in terms of the number of *characters* in error, the number of data *fields* in error, or the number of *records* that contain at least one error.

When we talk about a "character" we generally mean one of the 10 numeric or 26 alpha characters (i.e., numbers and letters), or any of a wide variety of special characters. A "field" is made up of one or more characters, and is generally an English word, a code or a command of some kind. One example of a field is your last name, another field is your telephone number, and still another field is your account number on your credit cards. Finally, a "record" is a form, card or CRT mask that usually contains more than one "field". Probably the best known example of a record is the keypunch card. Each card is a "record". Other examples include shipping forms, invoices, service orders, one page of typed copy, and one CRT screen full of data. One way to consider the three ways of reporting errors is that *records* contain *fields*, and *fields* contain *characters.*

Character-Level Error Rates

For character level error rates the number of characters in error is divided by the total number of characters entered. For example, if a key operator enters 10,000 characters on Tuesday and 150 of these characters are wrong, the character level error rate is calculated by dividing 150 by 10,000 and then multiplying by 100 (i.e., 150/10,000 x 100 = .15%).

Field-Level Error Rates

If we are interested in field-level error rates rather than "character-level" error rates then different numbers are involved. If the 10,000 characters represent 2000 fields (assuming an average of five characters per field), and the 150 errors occurred in 150 different fields (one error per field) Then the field-level error rate is calculated by dividing 150 by 2000 and then multiplying by 100 (150/2000 x 100 = 7.5%). There is obviously a big difference between the character-level error rate of .15% and the field-level error rate of 7.5% even though they both are calculated from the same error data. In fact, one is 50 times greater than the other.

Record-Level Error Rates

In most data systems, it is generally quite easy to tell when a character is wrong or a field is wrong, but when is a record considered to be wrong? Usually, a record is considered as being in error when one or more fields on the record are wrong. That is, an entire record is wrong if just one field on that record is wrong. The record may contain 20 different fields, and if just one of these fields is incorrect the record is wrong.

Thus, a third way of reporting on errors is the "record-level". Continuing with the same keying example, if the 2000 fields were equally distributed with 10 fields per record, there would be 200 records keyed. If each of the 150 errors occurred on a different record (one error per record) then the record-level error rate is calculated by dividing 150 by 200 and then multiplying by 100 (150/200 x 100 = 75%). In this example the record level error rate is 500 times greater than the character level and 10 times greater than the field level.

COMPARING ERROR RATES

Thus, using exactly the same data at least three different error rates can be calculated, with the record-level error rate being the highest, the field-level next and the character-level being the lowest. Field-level error rates are usually the most representative of what is actually happening in most systems and is preferred. Record-level error rates never tell you how many fields in a record were in error — there could be one, two, or more. Character-level rates tend to be deceptively low.

In reviewing past studies or reporting new information, it is essential to know the level at which an error rate is reported. There is obviously a large difference in the percentage that results from using any one of the three levels. It is really amazing what people have been able to get away with when reporting error rates. A supervisor in one system estimated the total errors from a small (10%) sample of the documents processed each day by each clerk. When a clerk keyed 2000 records in one day, this supervisor checked two hundred of them (10%) and identified the errors in the sample. But when calculating the error rate they used the daily input figure as the divisor (2000), and not the number of items sampled (200). Thus, if 20 errors were found in the sample the reported error rate was reported as a very low 1.0% (20/2000 x 100), instead of the correct 10.0% (20/200 x 100). On the surface it looked like they had an exceptionally reliable system (i.e. a record-level error rate of 1.0%); but they did not. It was not long before users of the system began to complain about the horrendous number of errors in the data base (Bailey, Blank and Walker, 1975).

ERROR RATIOS

Both error rates and *error ratios* may be useful in reporting error data. Meaningful discussions of human error are many times clouded because it is difficult, if not impossible, to determine exactly what an "error rate" represents. The error ratio reflects the average number of characters, fields or records that will probably be processed before an error is likely to occur. For example, the error ratio of 1:20 at the character level is read, "We can expect an average of one erroneous character for every twenty processed." Or the error ratio of 1:3000 at the field level is read, "We can expect an average of one wrong field for every 3000 processed."

The error ratio is determined by simply dividing the total number of entries (characters, fields or records) by the number of errors. For example, if a person types about 5000 words a day in preparing various documents, and makes about 10 errors a day, the error ratio is 1:500 (5000 ÷ 10 = 500). The typist would type an average of 500 words for every error that is made. Or if a key operator keys about 8000 records a day, and averages about 50 wrong records, the error ratio is 1:160 (8000 ÷ 50 = 160).

REPORTING RATES AND RATIOS

If all computer systems people would report error data using both error rates and error ratios, and would always state what level (character, field or record) the error rates represented, present communication problems concerning human error would be greatly reduced. We all have read articles or heard someone state, "Our system only has 2% error!" What does that mean? It should now be apparent that we cannot tell what it means.

One system designer developed a system that produced a field-level error rate of about 10.0%, i.e., an average of one error for every 10 fields, entered. His boss told him this was too much error and it must be reduced. So the designer made numerous adjustments to his system. Nothing he did reduced the field-level error rate. Finally, he found a solution to his dilemma: he began reporting *character-level* error rates. The reported error for the system dropped dramatically and his boss was happy.

One final warning on the use of error data. Even if we accurately report the error rates in systems, the "error rate" itself does not generally tell all we need to know about errors. We should be equally interested in the *types, sources* and *causes* of errors. The problem we frequently encounter is that an error rate is about all most people ever know or want to know about system errors. A similar situation was noted by the Little Prince in an English translation of Saint-Exupery's book of that name when he says:

> Grown-ups love figures. When you tell them that you have made a new friend, they never ask you any questions about essential matters. They never say to you, "What does his voice sound like? What games does he love best? Does he collect butterflies?" Instead, they demand: "How old is he? How many brothers has he? How much does he weigh? How much does his father make?" Only from these figures do they think they have learned anything about him.

Error rates are important, but so are other characteristics. The unfortunate problem with error rates is that they are usually fairly easy to calculate, but at the same time they are also very easy to misuse.

EXERCISES

The following inventory control information is entered into a computer for each person in a large corporation. The company has about 90,000 employees. The different types of pens, pencils, desks, chairs, etc., assigned to each person are recorded (and kept current) by a person interacting with a computer. The system is set up with one record per employee. A typical record for one employee is shown below:

FIELD NAMES	*TYPICAL CODE*	*NO. OF CHARACTERS*
Typing table	A32	3
Pen	4579	4
Stapler	S6239	5
Clipboard	CLP213	6
Telephone	TEL043B	7
Desk pad	PD1	3
Desk	D321	4
Chair	CH450	5
Typewriter	TYPE26	6
Book case	BC45M29	7

Answer the following questions using the above information:

1. If the record-level error rate in this system is 20%, what are the field-level and character-level error rates?
2. If the field-level error rate is 10%, what is the character-level error rate?
3. If the character-level error rate is 2%, what is the average record-level error rate?

3

ERROR SOURCES

INTRODUCTION

An "error source" in computer systems refers to the *place* or *location* in the information flow where an error occurs. It is generally assumed that information flows from an original point of "input", through people and computer processing, and finally exits a system at a point of "output". Errors may occur at numerous different places while the information is being processed by people and the computer.

There are three major sources of errors in most computer systems. These include errors that are "inherited" by the system, those made by people in the system, and those made by the computer. In most systems human errors occur primarily at three sources: during data preparation activities; while filling-out a form, either through handprinting or keying on to a terminal screen; and while rapidly entering data using key-to-card, key-to-tape or key-to-disc machines. Computer errors usually occur for two reasons: faulty software or an equipment/hardware malfunction. The six sources of errors are illustrated in Figure 3-1.

SYSTEM BOUNDARIES

Each of the error sources will be briefly discussed, but before doing so the concept of "system boundaries" (Holt, 1969; Stevenson, 1970) will be introduced.

The "system boundaries" concept is very useful when considering error sources. The boundaries around any system are artificial, and unfortunately are frequently determined quite haphazardly. But to

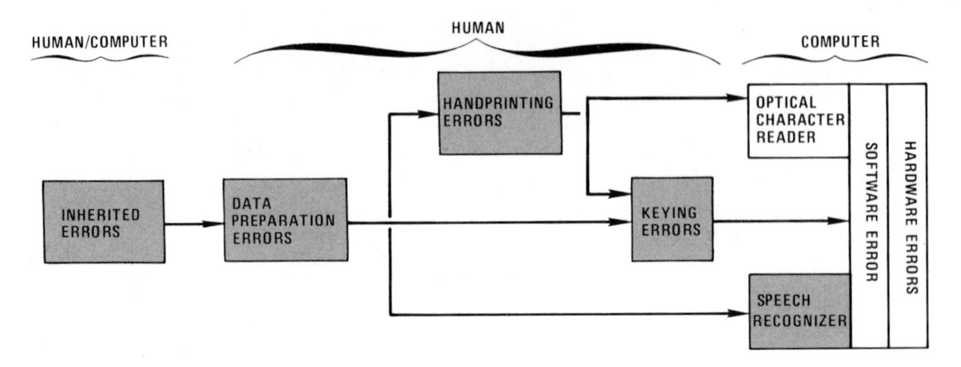

Figure 3-1. An illustration of six error sources in the flow of information in a typical computer system.

serious designers, system boundaries are a matter of major concern. The wider and more encompassing the boundaries are placed around a system, the more control the designer has over elements that are critical to the success of thc system.

Of particular interest to the discussion of error sources is the idea of widening system boundaries far enough away from the "paint on the computer" to encompass as much of the people-processing activities as possible. One of the gravest mistakes made by many system designers is concentrating solely on selecting an appropriate computer and creating programs to make it work, while excluding serious consideration of the activities to be performed by people who must interface with the software and hardware. A designer who is truly interested in meeting total system accuracy objectives will push the system boundaries out away from the computer as far as economically possible.

Figure 3-2 illustrates the added control over accuracy that can come from expanding system boundaries away from the point on the computer. Many systems are presently designed with only the computer and software in mind (Boundary A). Unfortunately, this approach dictates that the designer can only deal with software and hardware errors, and totally ignores the potential impact of all other errors. If the designer expands the system boundaries to include the transform module, only slightly more control over errors has been gained (Boundary B). This is because relatively few errors tend to occur at the transform source.

More control over accuracy can be gained by expanding the system boundary to include all transcription activities (Boundary C), but the greatest benefit to accuracy control comes from also including

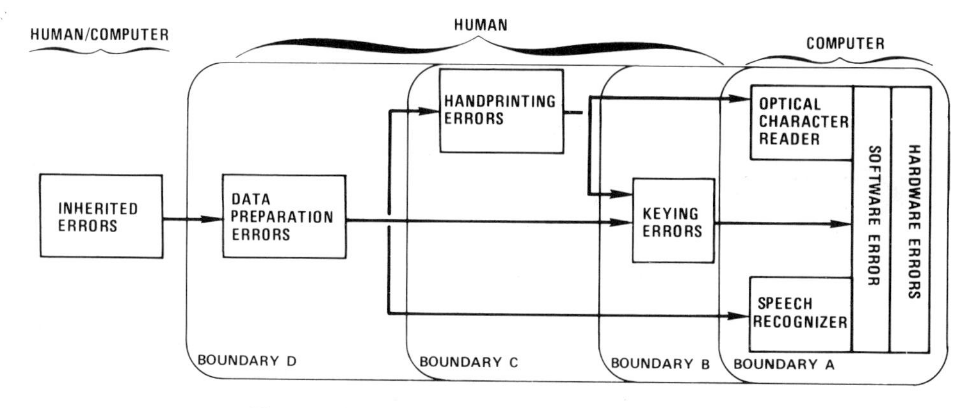

Figure 3-2. Illustration of four possible system boundary configurations.

all data preparation activities as well (Boundary D). Of course by expanding the boundaries a system designer assumes more responsibility for what will ultimately happen in the new system. But this also means that the designer will have more control over the number of errors that will occur. Given the opportunity, knowledgeable designers can develop systems that will minimize the occurrence of errors. Unfortunately, in the past few designers elected to take this option for increased accuracy control. Designers have tended to stay as close to the computer as possible. This left the people part of the system free to design itself, usually to evolve over a period of months or years. Times are changing and because of serious error problems being faced by many new systems, designers are now pushing system boundaries back further and assuming control over as much of the human performance as economically possible.

INHERITED ERRORS

All errors that are made outside the system boundaries, but which must flow into and through the new system, are considered to be "inherited" by that system (see Figure 3-3). If the system boundaries are set just outside the side panels of the computer, virtually *all* human error will be considered as "inherited" by the new system. A designer will have difficulty preventing any of it.

If the system boundaries include selection of an input device (after a human factors evaluation), there is slightly more control over errors, and if the system boundaries include the careful design of the human/computer interface, the selection and training of users, and preparation of instructions and performance aids, then the control over

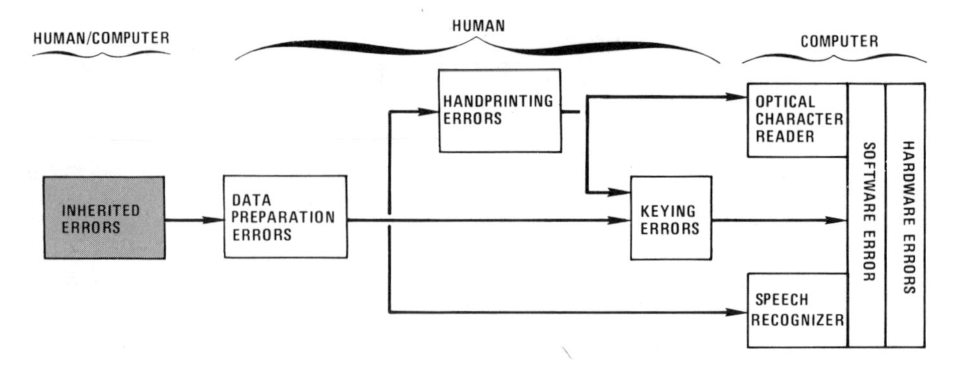

Figure 3-3. An illustration of several errors sources in computer systems with the "Inherited" source highlighted.

errors has been substantially increased. It is not difficult to see that by having wider system boundaries, a new system inherits fewer errors. The new system is the heir of fewer errors, because designers have more control over accuracy in the new system.

Inherited errors tend to be a major error source of errors in most systems. Because these are errors that already exist in the data when it is received, there is usually little a system designer can do to prevent these errors once the system boundaries have been established. The number of errors to be inherited by a system is determined by the cumulative total of all error sources from one, two or many systems outside the new system boundaries.

There is probably never a completely new and independent system. All systems have to depend on one or more other systems for information. As an example of what can happen in the inherited error process, assume that a new house is built in the country, the first house of many that will be built in a certain area. The correct address of the house is 7632 Chestnut Drive. The builder goes to the city offices, finds the address, and writes it down. In doing so, he transcribes 765_2 rather than 763_2 and abbreviates "Drive" to "Dr." He then contracts with a painter to put this number on the front curb for the new owners.

The painter interprets the builder's wrinkled, handwritten note as *1*652 Chestnut Dr., and in order to save paint, omits "Dr." When the new owners move into their house, they believe they live at 1652 Chestnut. Because their old address was "Sherwood Road," they add "Road" to the end of the address. Shortly after moving in, they call the telephone company business office to order a telephone and give their address as 1652 Chestnut Road. The telephone company service representative in taking down the order misspells Chestnut as Chesnut (omitting the first "t"). The information is entered into the telephone company's computer system as 1652 Chesnut.

This information is printed-out and given to the telephone installer who drives around for hours searching for the house. "That lousy computer system," he complains, "You just can't count on it." The computer system inherited errors from the new owner who inherited errors from the painter who inherited errors from the builder. It is probably safe to say that virtually all computer systems inherit errors from beyond their boundaries. The question is, "how much error is being inherited?"

If a new system is going to inherit a substantial number of errors, special steps must be taken to emphasize error detection in its design. Some inherited errors can be manually detected by proofreading inputs from outside the system boundaries. Other inherited errors can be detected by the computer. Once errors are detected, it is highly desirable, although not always practical, to send them back to the original error makers to be corrected.

One final thought on inherited errors. Some systems are "error-through" systems, while others are "error-end" systems. In an "error-through" system, an error can be inherited, pass undetected into its database and perhaps even pass to another system. The error may occur in the "error-through" system, but the wrong information is never used by that system. However, in an "error-end" system much emphasis is placed on detecting all errors that are either inherited or generated by the system. This is because the system must use the information, including all the errors inherited from other systems. A good example of an "error-end" system is the one used by the telephone company to prepare the white pages telephone book. An enormous amount of effort is made to detect and correct *all* errors, including all inherited errors.

If an "error-end" system is being designed, be prepared for having a substantial amount of human activity in correcting errors. It is a good idea to not try to sell an "error-end" system to your management using the argument that the new system will save them people, because it *will not.* In fact, if enough errors are detected, the new system may even need to *add* more people than the old manual system that it replaced.

DATA PREPARATION ERRORS

Once data enter a system boundaries (the new system boundaries will hopefully include many manual tasks besides simple data keying) people usually must do something to prepare it for entry into the computer. People frequently accumulate, summarize, classify, make preliminary calculations, prepare for routing, select relevant information, code certain material, and perform a variety of other tasks. In other words there are generally a large number of activities necessary to *prepare* data for entry into the computer. The "data preparation" error source is highlighted in Figure 3-4.

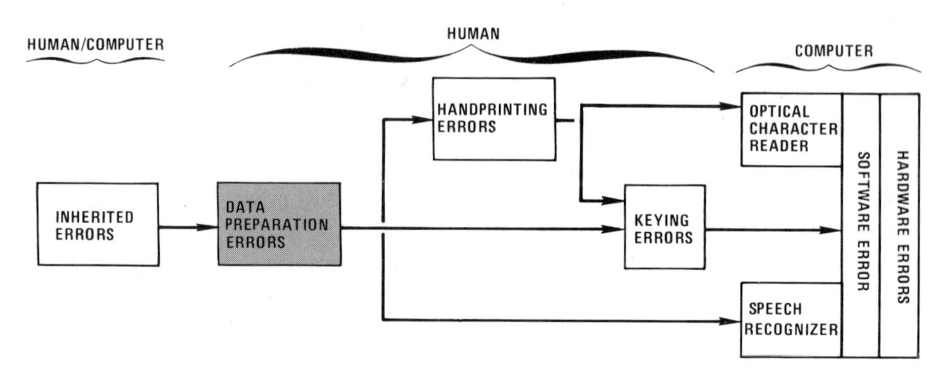

Figure 3-4. An illustration of several error sources in computer systems with the "Data Preparation" source highlighted.

The errors occurring at this source usually result from making wrong decisions, miscalculations, faulty judgements, etc. About four out of every five errors (about 80%) that will be made in a new system will be made here (Mitchell, 1979; Bailey, 1973; Manley, 1973). Because of the large proportion of errors that are made at this point in the information flow, it should be apparent that it is just as important for the designer to design the work for people, as it is to design programs for the computer. If the "data preparation" work is designed properly and adequate instructions, performance aids and training are provided, a designer can dramatically reduce human error in a new system.

It is curious that many system designers do not even try to reduce "data preparation" errors. They let these errors occur, slip into the computer, and then attempt a heroic effort to detect them. This obviously requires considerable more programming time for error detection routines, and once errors are detected, it usually requires much more time for people to make all the corrections. To make matters worse, computers do not detect very well the kinds of errors that people make in the "data preparation" source. Estimates are that only about 20% to 50% are usually detected (Bailey, 1973; Strub, 1975). However, the percent of computer-detected errors may be as high as

70% in some special situations (Bailey and Desaulniers, 1978). Nevertheless, it still makes sense for a system designer to try to prevent these errors from ever occurring in the first place.

TRANSCRIPTION ERRORS

Once appropriate decisions have been made, and all "data preparation" activities completed, the next step in most systems is to write down or key into a computer the results of the decisions. Errors made while "form-filling" either by handprinting or while keying are transcription errors. Keep in mind that each time a transcription activity is performed in the flow of data, a certain amount of error could occur. That is, each time a person is required to complete a handwritten form or enter data into a computer using a keyboard, an opportunity has been provided for errors to occur. Consequently, many system designers try to minimize the number of times the same piece of data is handled by people in a system. About 20% of all errors made within the system boundaries of a new data system come from the transcription error source.

Transcription errors are usually the result of transferring data intact from one source to another, without the requirement for making decisions while making the entries. This transfer of data may be from a telephone message to a form, a worksheet to a form, from one form to another form, from a form or worksheet to a CRT screen, etc. Transcription is basically a copying activity, using either handprinting or keying.

There has been a moderate amount of research conducted on transcription errors. We can use many of the conclusions from this research to help reduce errors in the design of new systems (the term "system" has been expanded here to include both people and computer processing). Even though there are many commonalities with errors made by handprinting and keying, there are also some rather obvious differences. Thus, each will be dealt with separately.

Handprinting

Most computer-based data processing systems are dependent in some way on handprinting (see Figure 3-5). It is amazing that the pencil, which dates back at least 2000 years, is still used side-by-side with modern computers. Try to visualize even the most up-to-date computer system without someone, somewhere using a pen or pencil while preparing or analyzing the information. Handprinting is still a very important activity in the operation of computer-based systems, and will probably remain so for some years to come.

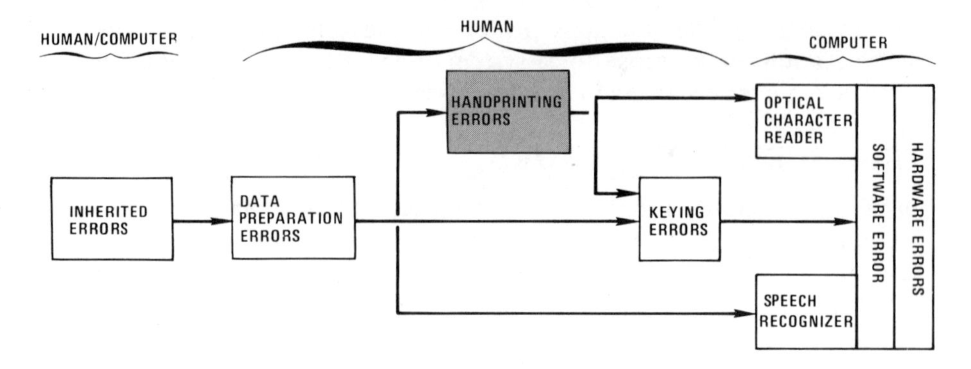

Figure 3-5. An illustration of several error sources in computer systems with the "Handprinting" source highlighted.

All handprinting requires an original stimulus of some type. The stimulus may be immediate and visual (copying from a list), it may be immediate and aural (printing a message being transmitted over the telephone), or it may be mentally generated by a clerk based on past learning (mentally converting a request for a color telephone to its appropriate code and then printing it on a form). No matter what the origin of the stimulus, in all cases one or more alpha characters (letters) or numeric characters (numbers) are copied on to a sheet of paper with the intent that they will be read by a person or a machine (OCR) at some later time.

Probably the biggest single problem with handprinting is that characters are sometimes printed illegibly and are consequently misread. Errors occur for other reasons too, but the vast majority of errors usually can be traced to illegible characters. Frequently characters are not totally illegible; they tend to be degraded just enough to look like another character. Some characters have a tendency to look more like other characters than do others. Appendix A shows the results of one study. Note that the letter "W" elicited the fewest errors, i.e., was confused with other characters *least* often.

Appendix A also shows quite clearly that the number of errors associated with some characters is greater than with other characters. In fact, less than one-third of all the characters contributed 62% of the errors. Collectively, the stimulus characters I, Z and G accounted for about 36% of all errors. The ten characters that were involved in the most errors (about 62%) are shown below:

I Z G V J 5 0 N 1 4

People frequently have trouble reading each other's handprinting, and tend to confuse certain characters. Optical character readers

(OCR's) have similar problems, and also may confuse characters. This is one reason why it is so difficult to have handprinted characters read directly by a machine, and also one reason why entering data using keyboards has become so popular.

Keying

Keyboards are still the most commonly used method for entering data (see Figure 3.6). Their use and design will be discussed more fully later. For our purposes now, you should be aware that people make about the same number of errors using a keyboard (in a form-filling task) as they do when handprinting. Thus, about the same number of errors are made no matter whether a person is filling out a paper form, or "filling out" a form on a CRT using a keyboard. The types of errors differ, but the error rates are about the same.

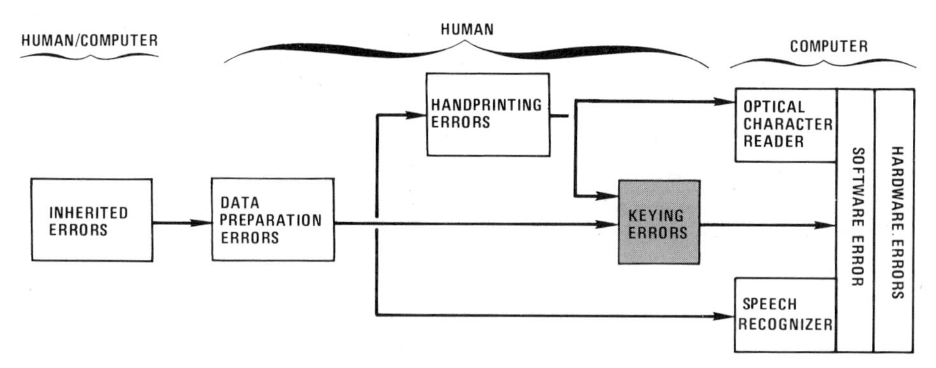

Figure 3-6. An illustration of several error sources in computer systems with the "Keying" source highlighted.

There are many types of CRT displays. Some are designed to be used like forms, where data are entered in blank spaces beside or below headers. Other displays are used with commands or computer prompts.

It is important to remember than when errors are made while either handprinting information on to a form, or keying information on to a large CRT screen, they are termed (for our purposes) *transcription* errors. In most data systems, transcription is usually done as a regular part of the daily routine, although it may be done only occasionally during the day. As much of this activity as possible should be within your new system boundaries, which means that design decisions can be made that will minimize the number of errors from this source.

One way that transcription errors can be reduced is by using carefully designed codes. The results of many studies suggest that most characters that are repeatedly substituted for one another are substituted because they have similar shapes, i.e., they tend to "look alike". Some characters tend to look more alike than others, which could be one explanation for a large number of substitution errors elicited by so very few character combinations. In fact, it appears that there is a direct and consistent relationship between the similarity of shape between two characters and the number of errors made for each *character combination.* As a result, there is an effective and very simple way of reducing errors caused by confusing characters, and that is to totally eliminate the characters that look alike. By eliminating certain alpha and numeric characters, a character vocabulary can be generated that substantially minimizes errors by reducing the use of "look alike" characters.

Using codes that are made up only of alpha characters or only numeric characters is a solution that has been used for many years. If a person in your system knows that a certain code is always made up of only numbers, there is little chance, for example, of confusing a "S" with a "5".

Based on our previous discussion, if a certain code is made up of all numbers (e.g., an account number such as 732468), this can usually help reduce some transcription errors. But in order for this to work the people who transcribe the code must be *aware* that it can be only numbers.

More will be said about how to develop codes that result in fewer errors later on. At this point in time, it is important to remember that both transcription and data preparation errors can be reduced by a knowledgeable designer who has control over these error sources.

Keep in mind that transcription errors usually only account for about 10% of *all errors* that enter the computer data base. Consider that about 50% of the errors will be inherited, and about 40% will occur during data preparation activities. Of course, as you make substantial efforts to reduce the errors in each of these error sources, you will probably change the ratios. You will probably be more successful in reducing errors in some sources than in others.

TRANSFORM ERRORS

Another error source is related to "transforming" data into machine readable form. These are errors made as a result of rapidly keying data from one source to another, without the requirement of making decisions. Transform errors are made while converting large volumes of data into machine-readable form by using a special input device, such as keypunch, a key-to-disc, or a key-to-tape device. This activity generally takes place in a special "keying area" and usually makes up the entire job of an employee. The main difference between transcription and transform activities is the strong emphasis on speed of entry and inputting large volumes of data daily in the latter case. In many offices these key operators will enter data at the rate of 14,000 keystrokes an hour (or more). The specialized nature of the transform activity, and the high skill level attained by key operators, results in substantially lower error rates.

The transform error source may or may not be in your system boundaries (i.e. under your design control). It depends on whether work is "sent out" for keying, or whether it is done by people inside your system. Obviously, you have more control over data accuracy if you have retained this activity within your system boundaries. But even if material is "sent out" for keying, the designer still has some control over the number of errors made. This is accomplished by designing well organized input forms, and providing highly legible copy for the key operators to work from.

SOFTWARE AND HARDWARE ERROR SOURCES

Our emphasis thus far has been on error sources related to the people operating within the system. It should also be noted that errors may occur at two computer error sources as well. The first source is bugs in the software, and the second has to do with computer hardware malfunctions. Errors related to computer hardware problems tend to be very rare, whereas errors in the software often range from excessive when first using a system, to minimal after several years of operation (see Figure 3.7).

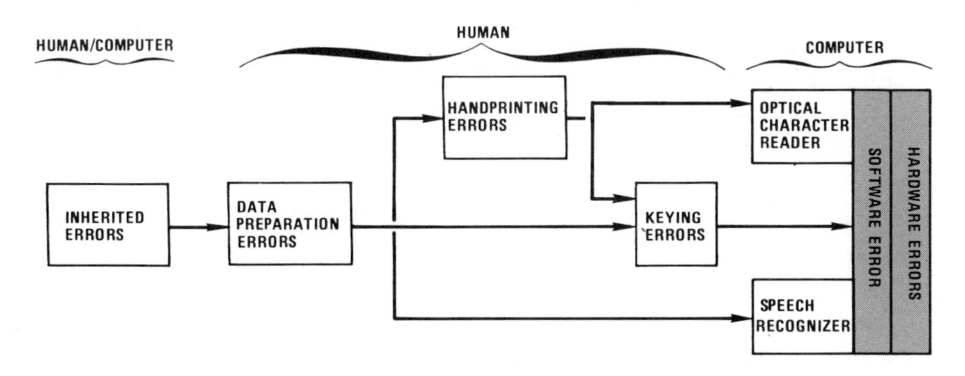

Figure 3-7. An illustration of several error sources in computer systems with the "Software and Hardware" source highlighted.

An example of a software error that did not show up until long after the new system was operational is when a high school senior honor student was stunned to receive a failing grade in English. A guidance counselor wrote on the bottom of his report card, "What happened?" The boy's mother wrote back to the school, "Ask your computer!" After checking into the source of this apparent error, school officials found that the boy had scored a perfect 100 in the course, but the computer was programmed to read only a *two digit number.* Instead of giving the boy credit for a 100, the computer had him listed as receiving a score of 10 for the course.

In another similar case, a 105 year-old grandmother received notice that she was to report to the local grade school to be registered. Again, the computer had been programmed to deal only with two characters. To the computer the grandmother was only five years old and it was time to begin the long educational process.

Most computer systems consist of both electronic elements and mechanical parts. The central processing unit, for example, consists almost entirely of electronic elements whereas most input/output equipment and file storage devices contain both electronic and mechanical components. Hardware failures can therefore result from the malfunction of electronic elements or mechanical parts. In addition, some failures can also be traced to faulty media, such as warped cards, magnetic tape with surface defects, and poor quality paper stock.

A COMPARISON OF ERROR SOURCES

If we compare the major error sources, we find that inherited errors can account for over 50 percent of the errors we must deal with in any given computer-based system. Data preparation activities also can produce a large proportion of errors. It is estimated that at least 40 percent of all errors come from manipulating the data (data preparation) before we ever write it down or enter it into the computer. Of course, these proportions could be much higher or lower in different situations.

It is possible to make some guesses as to "typical" error rates contributed by each human related error source. For comparison purposes some representative error rates for the four error sources we have just discussed is shown in Table 3-1. For example, the estimated error rate for the "transcription" error source at the "field level" and without any outside proofreading is about 2.5%. That is, we can expect that an average of 2 or 3 fields will be in error for every one-hundred fields we transcribe.

TABLE 3-1. Some representative error rates for four error sources.

	CHARACTER LEVEL		FIELD LEVEL[1]		RECORD LEVEL[2]	
	Error Rate	*Error Ratio*	*Error Rate*	*Error Ratio*	*Error Rate*	*Error Ratio*
A. Inherited[3] (Range)	2 - 5%	1:50 to 1:20	10 - 25%	1:10 to 1:4	87 - 99%	1:11 to 1:1
B. Data Preparation[4]	2.0%	1:50	10.0%	1:10	87.0%	1:1.1
C. Transcription						
a. No proofread[5]	0.5%	1:200	2.5%	1:40	39.5%	1:2.5
b. One proofread[6]	0.13%	1:769	0.6%	1:166	12.2%	1:8
c. Two proofreads[7]	0.06%	1:1666	0.3%	1:333	5.8%	1:17
D. Transform						
a. Key only[8]	0.03%	1:3333	0.15%	1:666	3.0%	1:33
b. Key & one[9] proofread	0.02%	1:5000	0.10%	1:1000	2.0%	1:50
c. Key & verify[10]	0.004%	1:25,000	0.02%	1:5000	0.4%	1:250

1. Assuming an average of five characters per field.
2. Assuming an average of 100 characters per record. The record level error data is derived from the character level data and assumes that errors occur as independent events. This is not to say, however, that all errors should be considered as being independent. Frequently in systems the occurrence of errors is completely dependent on the prior occurrence of other errors. Also to be considered is the relative distribution of errors throughout the records. A clustering of errors will result in a lower record level error rate because several errors could occur on the same record. On the other hand, a perfect distribution of errors, i.e., errors distributed evenly throughout all records, will result in a much higher record level error rate.
3. The inherited error rate is highly variable depending on the accuracy of the information coming from one or more other systems. These estimates were derived from telephone company studies.
4. Estimates derived from Cornell, 1968 and telephone studies.
5. Estimates derived from: Conrad, 1967; DeVoe, 1967; McArthur, 1965; Minor, 1962; Neisser, 1960; Smith, 1967; Tirrell and Klemmer, 1962.
6. Estimates derived from: Barone, 1965; Klemmer, 1959; Rabbitt, 1968; and Minton, 1969.
7. Estimates derived from: Barone, 1965 and Klemmer, 1959; and Minton, 1969.
8. Estimates derived from Elliot, 1959 and Klemmer, 1962.
9. Estimates derived from Barone, 1965.
10. Calculated from data contained in Barone, 1965 and Klemmer, 1962; derived from Minton, 1969.
11. Character level and record level error rates for inherited, data preparation and transcription error sources were derived from field level error rates. Field level and record level error rates for the "key" and "key and proofread" error sources were derived from character level error rates. The "key and verify" error sources were derived from record level error rates.

Consider another way to use Table 3-1. If someone were to ask you to predict what the error rate will be in your new system, you can add the error rates in Table 3-1 to come up with an estimate. For example, you may have found that your field-level would be about 10%, and add to this inherited error rate 10% for data preparation activities. If the data is transcribed on to a paper form (2.5%) and then transcribed on to a CRT using a keyboard (2.5%), you should add 5%

to the inherited and data preparation error rates. In this case you can estimate an overall field-level error rate of about 25%.

Inherited	10.0%
Data Preparation	10.0%
Transcription	2.5%
Transform	2.5%
	25.0%

Table 3-1 also shows some possible reductions by using *error detection* procedures. However, even greater reductions are possible if you design a system that prevents errors from occurring.

Keep in mind that error rates in Table 3-1 are estimates from studies that have been reported from a wide variety of different situations. Obviously, the actual error rate for each error source in your system will be determined by *YOU* as a system designer. As an informed designer, you can make decisions that will make these "typical" error rates either lower or higher in your system. If the error rates for each error source turn out to be substantially lower than those in Table 3-1, it will be due to some good design decisions on your part.

EXERCISES

Using the information in Table 3-1, determine the error rates for each of the following situations:

1. You have a friend who developed a new system that inherits a considerable quantity of errors from other sytems. The new system requires a data preparation activity, transcription with no proofreading, and keyed transforming with no proofreading or verification. What is the projected *field-level* error rate?
2. You have developed a new system that inherits few errors, requires one data preparation activity, requires transcription with two proofreads, and transforming that includes keying and verifying. What is the projected *field-level* error rate?
3. Assume that you have the same situation described in Question 2, but in addition, you have spent considerable time developing computer detection techniques. Your computer detection techniques are able to detect about half of the inherited and data preparation errors made in your system. What is the projected *field-level* error rate after computer detection?

4. Assume again that you have the exact same situation described in Question 2 (without any computer detection capabilities). Your supervisor is interested in impressing potential users by projecting the "lowest" possible error rate. What would be the projected *character-level* error rate?

5. Assume you are developing a system that inherits few errors, and requires no data preparation inside your system boundaries. Once information comes into your boundaries, a person reads it and then *transcribes* the information into a computer's database. There is no time for proofreading. Assuming no computer detection, what *field-level* error rate would you project?

4

ERROR PREVENTION

INTRODUCTION

Identifying where the errors occur in the information flow does not necessarily give the information we need to reduce these errors. What we really need to know is what things tend to *cause* errors to occur. That is, what can we do do produce a system that will have very few errors when being used. Obviously, to design a system in which few errors are made during its operation, a system designer must know what things are related to the occurrence of errors.

There seems to be a tendency to believe that making errors is a completely natural and inevitable part of being human. This is only partially true. There are usually definite and identifiable reasons why people commit errors; that is, most errors are *caused* to occur. If *all* causes could be identified and eliminated, we could have computer systems with perfect human performance. For this reason, the saying "To err is human" is nothing more than a frequently used excuse.

It is interesting that many people believe that there is a certain inherent reliability associated with performing certain tasks. For example, the reliability of the human in handprinting alpha-numeric codes has been suggested to be about 97.5% (Neisser and Weene, 1960). If true, this means that people will typically make about 2.5 field-level errors for every 100 codes entered (i.e., a field-level error rate of 2.5%). However, no matter what the so-called "basic" or typical human error rates suggested by this and other studies, the error rates can always be altered, either improved or degraded. Thus, depending on the skill and knowledge of the system designer, a system will have many or few errors.

CAUSAL FACTOR APPROACH

An approach to error prevention in data systems will be discussed in some detail. This approach is generally referred to as the "causal factor" technique, and makes four assumptions:

1. There is nothing inherent in manual activities (e.g., typing, encoding, keypunching) that will produce an error.
2. An error occurs when one or more "causal factors" act to degrade performance.
3. Many of "factors" that cause people to commit errors can be identified and controlled.
4. Under optimized conditions (i.e., with the causal factors controlled) people will perform activities with significantly improved accuracy.

Essentially, the "causal factor" approach suggests that if most error causes are identified and either eliminated or controlled, people performing in a new system will make few errors. This approach assumes that human generated errors can be prevented by reducing the factors that cause errors to occur. Unfortunately many system designers still feel that human errors are unavoidable and must occur as a natural consequence of doing work. But there is really *no* demonstrated lower limit on error rates in computer systems. System designers should strive to reduce the potential of making errors to the point necessary to insure optimum system efficiency. A good system designer will make decisions that keep error rates in a new system down to a very low level.

MAJOR FACTORS

Factors that have been shown to affect human error in computer systems can be sorted into seven major categories:

Design Factors
Written Instructions
Training
Environmental
Human/Computer Interface
Organizational Accuracy Requirements
Personal

Keep in mind that the terms inherited, data preparation, and transcription refer to error *sources* in the flow of data. Each represents a point at which the data is flowing through or coming into a system.

Within each of the error *sources* are the different *causes* effecting the accuracy of the data. Human errors result because of certain error causes, not error sources.

The seven major causal factors are shown on the following chart.

FACTORS RELATED TO ERRORS IN COMPUTER SYSTEMS

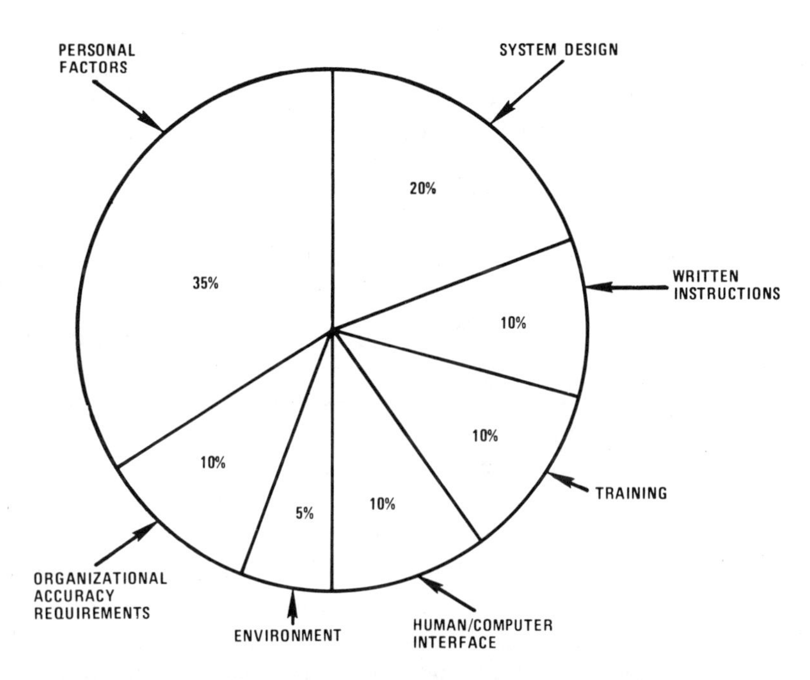

This type of chart is usually referred to as a "pie chart". The pieces of pie are not all the same size for a good reason. Some factors have a greater influence on errors in data systems than do others. The percentages shown on the chart are suggested to give you a feel for the relative importance of each factor.

One way to interpret the chart is to consider a computer-based system developed by a designer that had a good understanding of how to program and how a computer works, but had little idea of how people work. After the system was finished, people were hired to operate it. All the errors made by users over a one year period were collected and each was assigned to a causal category. Under these conditions, it is proposed that about 35% of the errors would be in the Personal Factors category, 20% in the Design category, 10% in the

Human/Computer Interface category, etc. Thus, different causal factors would be related to error occurrence more or less frequently.

A definite advantage of this causal factor concept is the ranking of factors and subfactors into an hierarchy of importance. Because of this added information concerning the importance of each factor, the system designer is better able to relate the relative criticality of each causal factor to system success. Past attempts at decreasing human error have tended to focus on only one or two of the causal factor categories, such as poor human/computer interface design, inadequate training, or some aspect of the environment. The result in many cases was a lack of significant or lasting improvement in preventing large numbers of errors. Including an indication of the relative importance of each causal factor category and related subfactors provides a structure for effectively addressing and reducing the number of human errors. This means that in order to have the greatest impact on the number of errors in your new systems, you should continuously emphasize those factors that are at the top of the hierarchy, i.e., those that have the most potential for degrading human performance.

Perhaps the most significant impact on accurate performance comes from personal factors. This is a consistent finding in many studies. Personal factors are probably responsible for at least 35% of the errors in most computer systems. Experience gained in developing, operating, and maintaining computer-based systems suggests that the design factor probably accounts for another 20% of human generated errors. Written instructions, training, supervision, and the human/computer interface account for another 40%.

In most computer systems, probably the least important factor is the environment. One reason why environmental factors are so well controlled in many computer systems is that substantial efforts are made to control the environment for the computer and its peripheral equipment. For example, high room temperatures (over 85°F-30°C) or low relative humidity (less than 20%) may degrade the reliability of the computer equipment. What is good for the computer is also good for people in the same area of the building. This causal factor category probably accounts for less than 5% of the errors in most computer systems.

RELATIONSHIP OF FACTORS

Table 4-1 illustrates through a series of examples the relationship of factors to one another, as well as the relative importance of each of the factors. The relative importance of the causal factors runs from left to

TABLE 4-1. A comparison of representative causal factors (subfactors)

POTENTIAL ORDER OF EFFECT	CAUSAL FACTORS					
	Personal Factors	*System Design Factors*	*Documentation Factors*	*Training Factors*	*Human/Computer Interface Factors*	*Environmental Factors*
1	Worker lacks a basic skill that is required, i.e., unable to read	Insufficient time allowed to complete an activity	Documentation cannot be understood by the workers	No training is conducted	Characters are not handprinted legibly	Lighting in the room is less than 10 foot candles
2	Worker is not interested in doing the job accurately	Manual activities are included that require a series of difficult and critical decisions	Manual instructions are not complete and/or accurate for all conditions	Training is not up-to-date	Extensive flicker or movement of characters on the CRT screen	The noise level is over 70 decibels with intermittent bursts to 100 db's
3	Worker is physically fatigued	No incentive for accurate performance is provided	Performance aids are not provided to assist memory	Formal training is not consistent with local practices	Excessive reflected glare on the CRT screen	The room ambient temperature is over 85°F
4	Worker has less than average short-term memory capacity	Manual activities are not grouped according to common skill levels	New manual instructions are not consistent with existing instructions	Training to update the worker on new information is not conducted	Codes are mixed alpha and numeric and over 8 characters in length (no grouping)	Inadequate workspace has been provided
5	Worker is over 40 years of age	Forms have been designed so that the location of fields on the form do not match the location of fields on the source records	Manual procedures contain numerous instructions that are poorly organized and formatted	No on-the-job follow-up of initial training	Orientation of the CRT screen is at an angle 70 degrees away from the operator's normal line of sight	The relative humidity is below 10%

right; the importance within each causal factor runs from top to bottom. For example, probably the single most important subfactor is "inadequate time to complete the activity", while the subfactor that would most likely have the least effect on the number of errors in a computer system is "low relative humidity".

The mere existence of one or more factors that are known to cause errors under some conditions does not mean that they will automatically cause errors under all conditions. The existence of "out-of-tolerance" causal factors only tends to increase the probability that errors will occur. The human is remarkably adaptive and has been known to work with a high level of accuracy under the most adverse conditions. Perhaps the only factors that almost guarantee that errors will occur are those on Table 4-1 that have the potential order of effect of "1" (first order), e.g., the office lighting is so dim that it is very difficult to see, or written (or verbal) instructions that cannot be understood, or insisting that a worker complete a data preparation activity in 30 minutes that realistically requires at least 90 minutes.

If two or more first order factors were affecting a worker at the same time, the resulting number of errors could be very high. On the other hand, it could be possible that all of the low order factors shown in the Table may be interacting and still result in few errors.

Ideally, all factors should be optimized in a new computer system, with the more-important factors considered first. A small adjustment to one or more of the most significant factors will have considerably more effect on reducing errors than optimizing several less important ones. For example, shortening the length of codes (less important) may have virtually no effect on system accuracy if one of the more important factors, e.g., inadequate training, has not been optimized first. Even so, the goal of a system designer should be to optimize all factors that may cause errors.

All factors cannot be optimized with a similar amount of time and effort. Perhaps the previous emphasis in many computer systems on optimizing many less important factors, at the expense of more important factors, resulted from the fact that some less important factors were relatively easy to optimize. Shortening the length of codes, for example, is much easier than optimizing job satisfaction. Thus, the relative payoff differs across factors just as the relative effort to optimize also differs. The tradeoff depends on the constraints (sometimes artificial) imposed on the system, such as available money, personnel, or training time. But keep in mind that in an attempt to reduce errors, some system designers only try to optimize the "easy" or most convenient error conditions. This approach could have the effect of only a slight impact on errors.

The causal factor categories have been presented as though they were mutually independent. Obviously they interact and are many times actually interrelated. That is, even though the room temperature rises to 95°F the user may not begin making errors until he or she also becomes fatigued. The fatigue and high temperature occurring separately may not result in errors, whereas together they may. In some cases fatigue and high temperatures may be acting on the worker and yet a strong desire to succeed (high motivation level) keeps the performance error-free. Because the several different factors are related to each other in numerous different ways, a system designer must consider what may happen if two or more major conditions are "out-of-tolerance" together (i.e., at the same time).

CRITICAL DATA FIELDS

One final note. Early in your system design, it is important to identify, organize and list all data items so you will be able to focus on those fields that are most critical to the success of your new system. This amounts to identifying the relative criticality of each data field. Errors in some data fields will have much less impact on total system performance than errors in other data fields. It is particularly important to know which data fields are critical to system success. As soon as you have defined every different data field to be included in your system processing, you should assign each a criticality level.

When trying to determine how critical a data field is to system success, one of your primary considerations should be the *consequences* of an error in that field. If an error occurs in that field, will the new system grind to a halt? If so, the data field is definitely a candidate to be called "critical". Few data fields are *that* critical, but many do have consequences that substantially disrupt or degrade system performance. For example, if your computerized data base (customer file) can be accessed only by inputting the correct account number, and the account number is wrong, you may not be able to find a customer's file. Or, if it is important to have absolutely correct money entries in order to collect money owed to you, these fields will be highly critical. The negative consequences to successful system performance of having erroneous account numbers or wrong sums of money indicates that these data fields should probably be considered as high critical.

It is convenient to divide all data fields into at least four categories: (a) highly critical to system success (b) moderately critical to system success, (c) low criticality to system success, and (d) simple through-put, i.e., carried by the system to but has no bearing on the success of the system. Thus, after you have a list of all the data fields your system must deal with, you should be able to go down the list and assign criticality. For example:

Data Fields	*Criticality*	*Reason*
Account number	Highly critical	Needed to find customer account
Customer name	Moderately critical	Good public relations
Customer address	Highly critical	Needed for mailing
Amount owed	Highly critical	Needed to bill customer
Number of children	Through-put	Possible future use
Number of cars	Low critical	Needed to send extra cards
American citizen	Through-put	Possible future use

The above data fields are from a fictitious oil company credit card data system. It should be obvious that not all fields deserve the same amount of attention when trying to prevent, detect and correct (i.e., control) errors.

EXERCISES

Assume you have been asked to study and analyze the errors being made in four different computer systems, System A, System B, System C and System D. You take a sample of 100 errors from each of the four systems, and spend considerable time determining the *primary* cause of each error. After establishing the most likely cause for the errors, you assign each error to one of the major causal factor categories.

Your results for each system are summaried in the following table.

Errors Assigned to Each Causal Factor
Category within Each System

Major Causal Category	Systems *A*	*B*	*C*	*D*
Personal	30	35	10	14
System Design	5	20	10	14
Documentation	30	10	5	14
Training	10	10	5	14
Human/Computer	10	10	60	14
Environment	5	5	5	13
Organizational Accuracy Req	10	10	5	14
TOTAL ERRORS	100	100	100	100

In System A, for example, of the 100 errors that you examined, you determined that 30 were caused by difficulties with Personal Factors, 5 by System Design problems, 30 by poor written documentation, etc. Compare the results for each system with the information for a *typical system* contained in the pie chart on page 35. Using this information, prepare a one-page report discussing the strengths and weaknesses of each system.

5

SYSTEM DESIGN FACTORS

INTRODUCTION

As discussed before, not all of the factors have the same effect on the number of errors made in your system. By working hard to optimize certain factors, the payoff is much greater than with others. One of the factors that if *not* optimized can lead to substantial errors in the "design factor". In fact, the "design factor", if not optimized, can contribute to as much as 20% of your system errors.

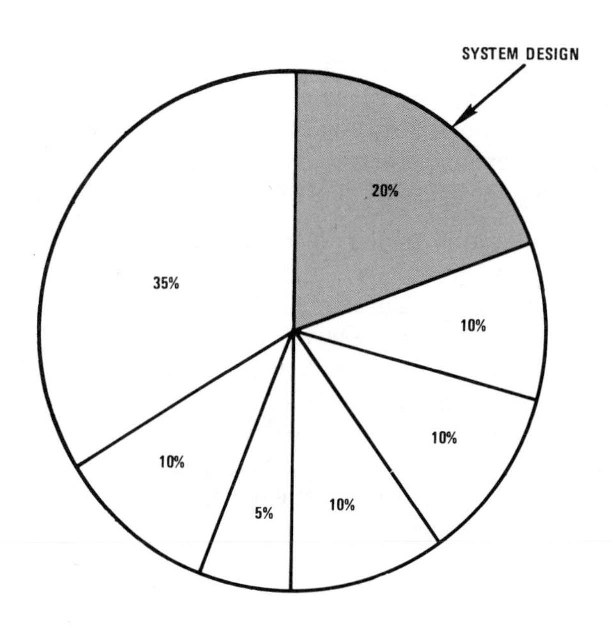

System Design factors refer to any overall system design condition that may ultimately reduce the reliability of the human component. This includes, for example, conditions which do not foster completion of manual activities in an acceptable time frame and with few errors. You could actually design a system that no one could efficiently operate. Many have done so! At any rate remember that System Design factors include any *overall* system level conditions that increase the probability of human error.

You should be very much aware that whatever you finally offer as a "system design" will be essentially "cast in concrete" as far as being able to make any changes that will substantially affect human performance. Early design decisions will be added to later decisions, and in the end they all will be intertwined so tightly that changes become very costly in terms of both time and money. It is possible to design a system that develops (during training) and maintains (during operation) behaviors that result in large numbers of errors. Keep in mind that one very important reason why good human performance considerations must be constantly made early in the development of a system is that it becomes so difficult to change early decisions, as new ones get added on top.

ALLOCATING FUNCTIONS

There is no question that computers do some things better than people, and that people do some things better than computers. The term "function allocation" is used to describe the process of assigning work to be done by your new system primarily to people or the computer. You should carefully allocate or assign all functions to be performed either by the computer or by people. Functions are allocated after determining all the work elements or functions that must be performed by your new system in order to meet your system objectives. The process of making these assignments is called function allocation.

Function allocation means to make the best distribution of work to the computer and to people. It is hardly ever practical to have the computer do everything. You should be very careful that you do not simply computerize what is easy to program, and then turn over to people all that's left. Too often in the past and even at the present time, people have been twisted to suit the needs or requirements of computer processing. A well designed system will always demonstrate a balanced distribution of work between people and the computer. Because we have so few firm guidelines, a set of well allocated functions usually illustrates the work of a knowledgeable and creative designer.

Thus, one of the most significant design decisions you will make pertains to which things you will have the computer do, which things you will have people do and which things you will have a continuous human-computer interaction in your system. Many "early-day" designers would look at all the things (functions) that had to be done, and then have the computer do those things that were the easiest to program. Whatever was left over was assigned to people in the system.

Mechanized systems are characterized by having almost all functions allocated to the computer, no matter whether this leads to the best system performance or not. However, good designers more carefully select the functions to be done by the computer, leading to *automated* office systems which are much more efficient than the old manual systems.

When trying to decide which functions will be allocated to people and which to the computer, keep in mind that people forget things fairly quickly. Do not have people do things that require constant or extensive use of their memory. In addition, people do not respond well to situations where they are either underloaded (have little to do) or overloaded (have too much to do). You are almost assured of many errors if you design a system that requires people to remember much and at the same time be overloaded with work.

A good function allocation will lead to fewer errors when your system is finally operating. This is primarily because people will *not* have been assigned functions, or work to do, that they typically do not do well. A good function allocation takes into consideration the *capabilities* and *limitations* of people and computers. A function allocation that takes into account the capabilities and limitations of both people and computers will lead to fewer errors when your new system is operational.

Your success in allocating functions to either the computer or people will depend, to a considerable extent, on your level of understanding about the capabilities and limitations of both computers and people. You should never start out by assuming that your system will require some people (i.e., you may be better off by computerizing everything) or that by computerizing some or all functions that things will get done better. It may be, for example, that all the functions that need to be performed within your system boundaries actually can be performed best by people. Nevertheless, when allocating functions to people or the computer, each function should be evaluated to determine if a person or the computer can do it best.

You should be particularly careful not to assign people to do things which will make excessive demands on them. It is usually best to exclude people from functions where:

(a) Perceptual requirements are near or beyond physiological limits,

(b) Responses will be physically difficult, conflict with established habit patterns, or cannot be readily checked or monitored for accuracy,

(c) Decisions require undue reliance on memory, require extensive organization of information, or must be made in too short a time to be practical,

(d) Communication requirements interfere with other activities, or

(e) Tasks may overload the human, resulting in inadequate workload distribution, or preclude adequate and timely monitoring of system status.

When allocating functions, always watch to make sure you have not "overloaded" any of the people in your system with work requiring several consecutive critical decisions. If you have overloaded any of your people with the kinds of activities that people don't do well, many errors will be made. There is also some evidence that if you "underload" people, they will soon become bored with the simple-minded activities you are expecting of them, and this situation may also lead to increased errors.

Also remember that occasionally the best solution for fulfilling certain functional requirements is a people-computer combination (i.e., interaction) in which each complements the other in an interacting dialogue.

One final word about function allocation. In the past, probably the most used reasons for allocating work to people versus a computer were cost and availability. A computer which could perform all the functions that people could perform in a given system may well have cost more than 10 times as much to operate. But times are changing. People (and paper) are becoming more expensive and computers are becoming less expensive. The question of whether to use a human or a computer in a given function only can be answered after consideration of many factors, including the possible tradeoffs which produce maximum efficiency in terms of *cost.*

PROVIDING SUFFICIENT TIME

Once you know which functions are to be performed exclusively by people and those that will require human-computer interaction in your new system, you are ready to make several more design decisions that will effect human performance. Another critical decision has to do with the *time* allowed for completing each manual activity. You must constantly check and double check to make sure you provide a sufficient amount of time for users in your system to complete each activity. In a batch system this is not so critical, and may even be beyond your control. But in an on-line interactive system, you may feel a slight tendency to "hurry things up" by trying to shorten the time allotted for people processing. If people feel hurried or rushed, this can lead to numerous errors. One condition that will almost certainly lead to increased errors in your new system is to force people to perform faster than they are accustomed.

FEEDBACK

Another important "system level" decision that will affect the number of errors in your system has to do with *feedback*. For example, it is essential that a person become aware of the number and types of errors he or she is making. The sooner this feedback is given, the more effective it will be. As part of your system design you should make provision for feedback to come to each user from at least four places: data codes, human-computer communications, system reports and supervision. Providing system operators with information on the number and types of errors they are making can be very helpful in reducing future errors.

The source of feedback that is more frequently ignored by system designers is from "data codes". Only a brief introduction to the use of this and all other feedback sources is possible here. When a person handprints, keys or in any way enters a code, characteristics of the code itself should have the capacity to at least partially indicate whether it is right or wrong. For example, in America almost all telephone numbers contain seven numbers. So when you write or enter a telephone number and then review what you have written, if there are only *six characters* or if one of the characters is a letter, you know there is a mistake. The code itself helps to provide feedback about errors. As noted before the more immediate the feedback, the better. In the above example, the immediacy of the feedback not only informed the user that an error had occurred, but also provided him or her with the opportunity to make a quick correction. Thus, the lowest (first) effective level of feedback in a data system can come from the data codes themselves.

A second source of designer-controlled feedback occurs primarily in on-line, interactive systems and is called "human-computer communication" feedback. Human-computer communications feedback refers to information about an operator's performance that is detected by either the computer or the user. For example, a user tries to call up a mask to enter some new data on to a CRT screen. He or she enters what is thought to be the correct command, but the computer does not respond, or even better, sends back an appropriate error message. If the system designer is particularly innovative, the computer will not only indicate that the command was not appropriate, but will also supply a list of potential right responses (i.e., acceptable commands). Obviously, the system designer must anticipate the need for feedback, and make the necessary provisions early in the design process.

The last two feedback sources have less effect on error control and will be discussed together. Feedback from system reports (i.e., daily, weekly or monthly printouts) is usually too general and too late to be of much help to an operator. To tell users at the end of the day or at the end of the week that they made 78 errors has little meaning. Operators need to know exactly what their errors were, and what correct responses should have been made. Along the same lines, supervisors tend to give the same general type of (limited value) feedback as do many computer reports (frequently, their only source of information is the computer reports).

As one philosopher has said, "The greatest of faults is to be conscious of none." Even though there are many ways of dealing with feedback, you should try to incorporate methods that will give the best results. One study reports a tremendous reduction in the number of errors being made by workers by using feedback produced by a large GMC truck horn. Every time an error was made and detected by the input device, the truck horn would blow behind the worker. The reduction in errors was phenomenal. The workers knew exactly what was acceptable. This is probably an extreme solution for most computer systems, but it certainly makes the point that adequate and meaningful feedback is a must.

We have been talking primarily about feedback that told users what and when they did something wrong. You should also make every effort to give people feedback when they perform like they are supposed to (i.e., do things right). This can be done by providing periodic computer messages like "Very good, you have not made an error in the last 100 entries!" Your goal should be to design software that "glows after good" and "growls after bad" (Mager and Pipe, 1970). It is well within the state of computing technology to provide this type of feedback. Let your computer evaluate each operator's performance.

INTEGRATING MAJOR COMPONENTS

It is not enough to think only of each part or component separately, or even to try to optimize each part separately. A good system designer must carefully and skillfully *integrate* the people and computer components. That is, we first separate all the functions, assign each one to either people or the computer, and then try hard to bring about a workable integration or "working together" of the parts. An example of the need for integration can be seen in our use of the automobile. We have many good cars, roads and road signs. But occasionally, when driving in a hurry, we find that we come to an unfamiliar turn in the road, and when trying to make the turn (at the speed designated by a sign) our car drifts into the lane of oncoming traffic. An accident occurs, and we look for the cause. Is it the car? the road? the road sign? Probably the best answer is that it was the faulty *integration* of the three together. Separately they were fine, but the way they came together on that particular day was not good design.

DESIGNING WORK

One final "design factor" condition deserves to be mentioned. Human performance experts have found that by having "easy" tasks and "difficult" tasks mixed together in some unsystematic way, that it tends to increase the probability of errors being made. One of your goals should be to try to group easy tasks together, and more difficult tasks together. Don't create a situation where you have a person performing a couple of easy things, then a difficult item, and then a very easy item, and then two extremely difficult items, etc.

Try to organize your work activities into small work modules. For example, provide one or more work modules that include only the less difficult tasks that a person (someone recently hired) could be easily trained to perform. Provide other work modules that contain more difficult tasks where a second person (one with more experience) could be trained. If each work module contained tasks that could be easily learned and performed by users at different skill levels, you stand a much better chance of matching available skills to appropriate tasks. Briefly stated, a good system designer will develop a system that has all manual activities systematically grouped into *modules* with common skill level requirements.

EXERCISES

The purpose of this exercise is to demonstrate the effect of visual feedback on self-detecting errors. You will need four typists to help with this exercise (you can use yourself as one of the typists). Have each person type two different selections, each lasting exactly 5 minutes. Selection A begins in this book on page 1. Selection B begins on page 65. Have two of your people type Selection A first and Selection B second, and the other two type Selection B first and Selection A second.

Tell each person to type "as quickly and accurately" as possible. Have each person type "qq" after each error they make and detect. The typists are to make no attempt to correct their errors, or to retype a word. Their main goal is to type as many words as possible during the five minute period, and to indicate when they make errors.

The people can use either a typewriter or a video display terminal (keyboard and CRT). When they type Selection B *do not allow them to see what they have typed.* With a typewriter, cover the paper being typed, and with a video display terminal cover the CRT.

After all selections have been typed, count the total number of errors made, and the number of errors self-detected by all typists. What percent of errors was self-detected? To find out, divide the number of self-detections by the total number of errors made, and multiply by 100.

Write a one page report discussing your results. Did the ability to see the material being typed (i.e., visual feedback) have an effect on the number of self-detected errors? Did it have an effect on the total number of errors made? Is visual feedback important in this type of task?

6

WRITTEN INSTRUCTION

INTRODUCTION

Written instructions have a potential impact large enough to be of considerable importance during system design.

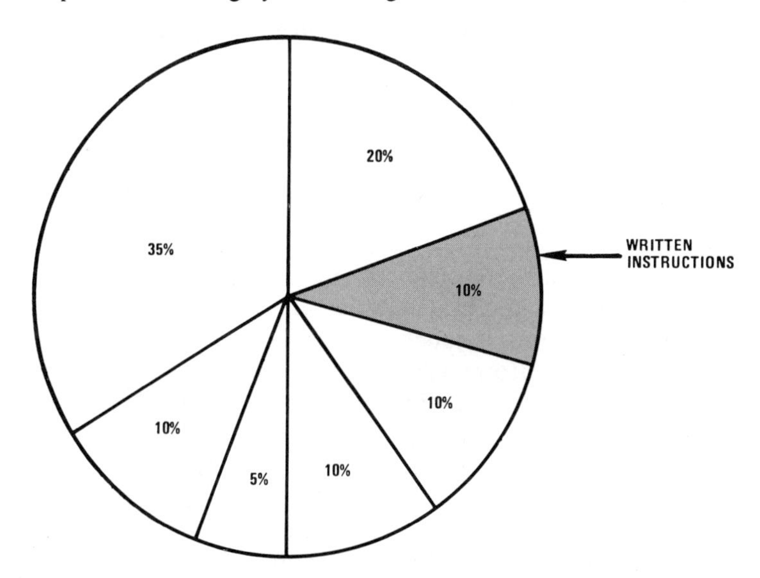

You may have wondered how many different documents need to be developed per system. This, of course, depends a great deal on the nature and size of the system. At least three types of documents may be developed: (a) a system design specification, (b) computer subsystem documents, and (c) a good set of written instructions. The "system design specification" contains a detailed listing of the system parameters, including capabilities, inputs, outputs, specifications, etc. This document may never be seen by the people who will be operating or using your system. The computer subsystem documents include program descriptions, data file descriptions, program listings, and run books. Only the "run books" will be used to operate your new system, which means they should certainly meet the requirements of being complete, clear and up-to-date.

Complete, clear and up-to-date instructions are required for terminal operators, error corrections clerks, data preparation clerks, etc. That is, any person who operates within your system boundaries deserves to have a document that tells him or her exactly what they must do in order for the system to meet its objectives.

Someone has said that "the next-best thing to knowing something is knowing where to find it." With each system we find that people using the system have numerous questions about how it operates, exactly what they are to do, when are they to do certain things, etc. In order to best meet the needs of system operators and users, you must provide a good set of instructions. System users simply are not able to remember all that they learn in training, and even those few that do seem to remember will soon be promoted, quit or in some way find another job.

PERFORMANCE AIDS

Some instructions should be in the form of *performance aids.* Performance aids are documents that store informations for *ready use* by a system operator while he or she is performing an activity. Performance aids are distinct from other system documents or training materials in that performance aids are designed to be used "on demand" while performing a job. Of all the documents you will produce, probably the performance aids will have the greatest impact on the number of errors made in your system. Well developed performance aids may reduce training requirements and the need for extensive procedural documentation. At the same time, good performance aids will lead to improved job performance (i.e., fewer errors). One of the simplest examples of a performance aid is the list of frequently called numbers you have by your telephone. This list saves you time and results in fewer errors than if you had to look up every number in a large telephone book, or even worse, if you tried to remember all those numbers.

You should develop performance aids whenever and wherever you have a situation that places a heavy burden on an operator's memory. You don't want them spending hours paging through lengthy documents, nor do you want them to reach the point of habitual guessing. Think ahead to situations where a "memory aid" may be useful, and then develop them. After your system is operational, observe to see if there is a situation that you overlooked that requires a performance aid. You will frequently find that shortly after your system is operational many operators will have already developed crude and sketchy performance aids to help them do the work better. You should be somewhat uncomfortable if this occurs too often, because it indicates that you did not consider both computer and human performance in your design.

PREPARING INSTRUCTIONS

Remember, people who do not have good written instructions either ask someone else what to do or guess. The person they ask may or may not know the correct answer, and obviously many guesses on how to proceed will be wrong. Thus lack of systematically prepared instruction could very easily lead to increased errors. But not only must the necessary instructions exist, they also must be prepared in a way and at a level that *each* intended reader can easily understand them. A good system designer will always prepare *complete, clear* and *accurate* instructions for users.

Instructions should be written with sufficient detail and at a level necessary to communicate to users. One way to determine the "readability" level of documents is to apply a "readability index" like that shown in Appendix B. This index provides a fairly easy way of determining at what "grade level" or reading level your documents are written. Once you have determined that your instructions are *complete* (i.e., all conditions are covered), *clear* (i.e., you have included little or no jargon, you have written on a level that each reader can understand, etc.), and *accurate* (i.e., the written word accurately reflects the way the system really operates), you are ready to apply the readability index.

There have been numerous studies relating the readability indexes to human performance. One study (Kulp, 1974) found that documents having a computed grade level of 10.0 or higher tended to have a negative impact on human performance where people were trying to understand printed instructions. Another investigator (Sticht, 1970) found that the greater the difference between readability level of printed material and the ability of the reader, the *less* the material tends to be used. So in order to help ensure that your instructions are used, and that the material is clearly readable, it is not a bad idea to keep the readability grade level at 10 or less.

The problems associated with having poor instructions are just beginning to be systematically studied. Even so, time and time again, it has been observed that a major cause of error is the fact that users have little or no idea what they are supposed to do. In some systems, a user's responsibilities may never have been written down, and if the person who previously had the job forgot to tell the new person something, the user is likely to perform some activities incorrectly and make numerous errors over a long period of time. In fact, the exact same types of errors may continue undetected for weeks, months or even years.

There's a curious myth about some technical writing, including the preparation of instructions. Many people believe that difficult or obscure writing is the mark of a wise and learned system designer. In actual fact the reverse is true. Anyone can be obscure and incoherent. Mental hospitals are full of such people. But to clearly write instructions takes real skill. As a designer you carry the responsibility to develop efficient *software* that computers can understand and quality written instructions that people can readily understand.

A good example of poor instruction is illustrated by a problem that arose in a very large building in Baltimore (Chapanis, 1964). Someone had a sign hung between two elevators that read:

PLEASE
WALK UP ONE FLOOR
WALK DOWN TWO FLOORS
FOR IMPROVED ELEVATOR SERVICE

It was discovered that most people thought something like, "This must be one of those fancy new elevators that has something automatic. The elevator doesn't stop at this floor very often. If I want to get the elevator I'd better go up one floor or go down two floors." And this was what people did! Unfortunately, after trudging either up or down, they found exactly the same sign again.

What the people who developed this seemingly simple instruction were trying to say is shown below:

IF YOU ARE ONLY GOING
UP ONE FLOOR OR
DOWN TWO FLOORS
PLEASE WALK
(IF YOU DO THAT WE'LL ALL
HAVE BETTER ELEVATOR SERVICE)

It is difficult to always write what we desire to say. If your programs don't communicate, the computer will not work right, and if your written instructions are unclear, people will make more errors.

SPECIFIC ACTIONS REQUIRED

While you are preparing instructions, you should continually ask yourself, "What kind of specific human action will a person take when he or she reads this instruction?" Keep in mind that it is possible for instructions to be bad not only because they are obscure or confusing, but also because they are too brief. If you are in doubt about the ability of your instructions to adequately communicate, *test* them. Have people read them and do what they say without any help from you. Observe and take notes. Then revise your instructions until you are confident of the outcome. One important caution for such a study. Make sure the people you use for testing have about the same knowledge, skills and attitudes as those who will be using your instructions in the "real world". This point deserves repeating. It is usually very difficult to know if other people will truly understand our written instructions. For this reason, it is always a good idea to conduct tests on the material.

TYPICAL USERS

As you describe the work to be done, keep in mind the experience level of the person who will be reading the material. Many users have a high school education (or less), no courses in computer processing, and probably only moderate interest in being part of your new system. Keep a picture of this person in mind all the time you are writing. Don't be so foolish as to think that the people operating within your system boundaries will be highly educated, highly experienced or highly enthused about doing what needs to be done. Designers who plan on such super-people being available are almost always surprised and disappointed. Again, when developing instructions, try to keep a picture of the "typical" reader in mind.

NUMBERED LISTS

Many times written procedures communicate best if they are documented as a series of relatively brief instructions. Each instruction should be numbered and the instructions should be arranged in the order used. Try not to include unnecessary detail, but make sure you include enough information so that the "typical" reader will understand *exactly* what needs to be done. Remember that you are not going to always be around to answer questions. Your system users should be able to read and find *all* their answers. To help encourage the use of your instructions, why not include table of contents, indexes, glossaries, etc.

COMPUTER-BASED INSTRUCTIONS

A final word on instructions. With the cost of people and paper going up, and the cost of computer processing going down, it is now practical in many systems to store large amounts of user instructions in the computer. You should try to have computer-based instructions whenever possible.

Many computer users are seated at a terminal. A convenient way to access computer-based instructions is to use a "help" key. When the user is uncertain about how to proceed he or she presses "help". The material that would ordinarily be in a paper performance aid flashes on the CRT screen, or perhaps just one corner of the screen. The user then reads the instructions, and either proceeds or hits the "help" key again. If pressed a second time, the "help" key generates a more detailed description; and if pressed a third time, the user may get the training materials associated with the instructions, along with examples and some exercises. This approach reduces the need for paper documents, and if well designed, may lead to less manual processing time and fewer errors.

EXERCISES

Prepare a one-page set of instructions explaining to someone with *no computer experience* how to log onto the computer system you use most often. Also, instruct them on how to access the text editor, how to insert the short paragraph of text shown below, and how to replace one character with another.

Find someone with no computer experience, and have them follow your written instructions. Observe what they do, but *do not* give any assistance. If the person is unable to accomplish the tasks without help from you, revise your instructions and try again with another person. Prepare a one-page report on your experiences in providing a complete, clear and accurate set of written instructions.

Paragraph to be Entered

Using a text editor is fun once you figure out how it works. It is particularly helpful in making corrections to certain types of errors.

7

TRAINING FACTORS

INTRODUCTION

Training factors probably contribute about the same proportion of errors as do the written instructions. Both the training and written instruction factors contribute less than the design factors, but their potential for producing errors is still substantial.

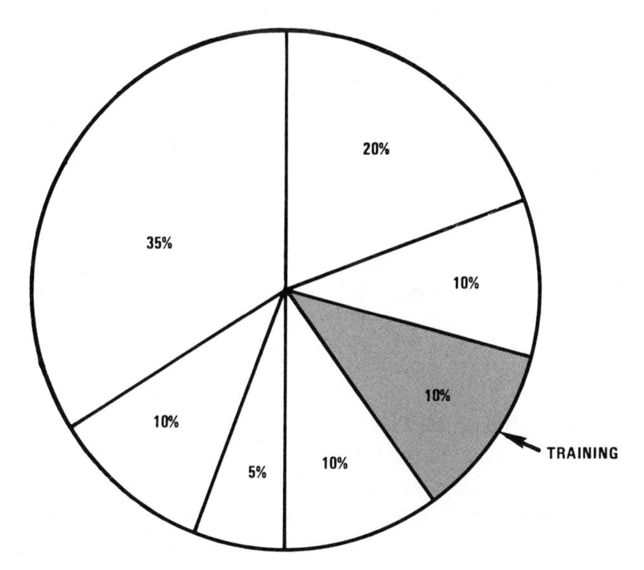

Providing inadequate training can be very costly. Back in 1974, the Social Security Administration came up with an answer to the welfare problems of 4.2 million blind, aged or disabled adults. They decided to use what was then one of the worlds largest computer systems. Unfortunately, they later found that during the first two years the computer-based system had sent out somewhere between $400 million and $1 billion in overpayments. One of the major reasons given by Social Security officials for these costly errors was not inadequate software or a faulty computer, but "poorly trained personnel."

A system may be well designed and have adequate written instructions, but if users are not taught exactly what they are to do and how best to do it, then the probability of having errors is high.

IMPROVING UNDERSTANDING

Unfortunately, users frequently perform activities under one of the following conditions:

a. users do not understand the activities they are to perform and know that they do not understand.

b. users do not understand the activities, but believe they *do* understand.

If a user does not understand and knows it, he or she may try to find out what to do. But if a person does not understand, but believes he or she does understand, then the same type of errors can be made for days, weeks or months before they are finally stopped. After training, the worst possible condition is where a user feels that he or she knows what to do, but really doesn't.

TYPES OF TRAINING

Acquiring all the knowledge and skill necessary to work efficiently and accurately may be difficult when only one or two short formal training sessions are used. Consequently, both formal "classroom" training and on-the-job coaching are usually required to reach the desired level of proficiency. Frequently, system designers do not take the time to develop even a "sketchy" training course to be used by new system personnel. In this case, the new users have no one to learn from except, possibly, the experience of those who worked in the system before them. Almost every time someone tries to learn from the experience of another, some things are left out and others are told wrong. Each time a new person is trained by another user the probability of new errors increases. By carefully developing training materials for each new system, a designer has much more control over what people learn to do in the new system.

It is essential when learning to perform new activities that error producing behaviors do not develop early and become habitual. This is why on-the-job training is most effective *after* a user has more formally learned exactly what needs to be done. We don't want new users to learn "wrong things" because these wrong behaviors have a way of persisting. If they learn right the first time, they have a better chance of doing things right from then on. Thus, it is important that new system users learn exactly what they are to do in an initial training course. This is so that first-learned habits are not error producing habits.

Even after a formal training course is completed, people usually require on-the-job coaching for a period of time. To do *all* training "on-the-job" is very inefficient and costly. You can imagine how inefficient it would be if a college football team learned all they needed to know only while playing their games. Coaching during the games is good for making minor adjustments to performance, but *not* to learn the "basics." We should learn how to operate a new system on the "practice field," and not on the playing field in front of 100,000 fans. So keep in mind that learning how to operate a system should take place in at least two different situations: in a formal course, and with on-the-job coaching.

LEARNING PRINCIPLES

Psychologists have been studying how people learn for over 100 years. A set of principles that should be considered when developing training materials was complied from this research (Hilgrad and Bowers, 1975). These principles are listed below. They contain some very useful advice for system designers interested in providing effective training courses.

1. Keep the trainees *active* (skill can be best developed by doing, not just listening or reading).
2. Make use of *repetition* (practice makes perfect).
3. Make use of *reinforcement* (reward correct responses).
4. Have trainees *practice* in many different situations so that they are able to generalize.
5. *Organize* the presentation of information in some meaningful way.
6. Provide for learning with *understanding.*
7. Encourage *divergent thinking* (urge students to develop creative solutions and explore alternative solutions).

8. Consider the trainee's *ability* to *learn* (some people learn quickly, some learn slowly).

CENTRALIZED TRAINING

If you are involved in developing a large system for a large company, there is another situation you should be aware of. In one company the users were trained in a headquarters office. When they went back to their offices in outlying areas, they found that many of the rules and conditions were not the same as they had learned during training. Each local area had its own local customs and procedures. The newly trained users had to come to a *new* understanding of what was to be done; that is, they had to unlearn and relearn. Relearning takes time, and during the relearning experience, there exists the probability of a tremendous amount of error occurring. It was not long before the outlying areas became very hesitant about sending people to the centralized training courses. People do not like to be trained on the "same" activity twice. We should develop training that takes into account exactly how things are to be done.

TRAINING/TESTING DATABASE

The best way to know if someone has learned all they need to know to be part of your system is to give them a test. Not a paper and pencil test, but a test using a terminal that actually exercises both the new user and the computer. Many system designers have developed special training/testing databases that allow them to see if newly trained users can really do what needs to be done. Samples of the most frequent and most difficult types of problems are given to each new user, who then performs a series of transactions. The advantages of a special testing/training database it that (a) a person knows in advance what the right response should be, so it is easy to detect errors and (b) the database can be easily returned to it's original form after each training session. If a new user meets some predetermined performance objective, say 98% of the items done correctly, then he or she is ready to perform on the job.

DEVELOPING TRAINING

Training materials should be developed *after* written instructions, including performance aids have been developed. And, in fact, should be written to supplement these items, not replace them. It is important to write down *objectives* for each training module, and then *produce materials* to meet those objectives. It's even a good idea to make up *exercises* for people to practice on, and then have a "*final exam*" when they have completed the training courses. Some training tests can be "paper and pencil", but the best tests make use of the same tools a person will use on the job (e.g., keyboard, CRT, etc.). If it is computer-based training, then most, if not all, training materials can be stored in the computer's data base.

No matter what approach is selected every system should have a set of carefully developed training materials so that new system users can learn their duties quickly and correctly, which will keep to a minimum the number of errors.

COMPUTER-BASED TRAINING

Computer-based training is frequently a good choice for new systems. A creative system designer should be able to make much use of the computer-based training techniques that have been developed over the past few years. Before using this type of approach, however, make sure it is being done for the right reason. Computer-based training should not be used because it is novel, neat or convenient, but because it is the "best way" to accomplish the learning of skills. A system designer should have as an objective to train people as *quickly* and *efficiently* as possible. In some systems, paper-oriented documents may satisfy this requirement much easier than using the computer.

One advantage of using the computer to assist in training people is that it not only can provide a very flexible presentation of materials to the learner, but it can keep track of the progress of the learner (in fact, several learners at once). The computer can be used to provide drill or practice, it can provide learning in the form of games or case problems, and it can be used to automatically supply remedial information at appropriate times. Good computer-based training can be very effective in reducing errors in data systems. In fact, one of the main uses of computer-based training is to provide practice with immediate feedback so the learner always knows if an error has been made and can correct it.

SPECIAL TRAINING NEEDS

Finally, it's a good idea to think ahead to a period of time about six months after a new system becomes operational. If the new system has some critical areas to be performed only occasionally by people, there may be special training problems. Consider for a moment the way airlines train their pilots for emergency situations. Like some tasks in data systems, airline emergencies (fortunately) do not happen very often. So "refresher" training must be provided to pilots every six months, so they can practice and remember what they are to do when an emergency arises. Each system designers should consider supplying a special "refresher" course as part of a new system. This course should deal mainly with items or conditions that are fairly critical, but occur infrequently. People need to be reminded of correct procedures from time to time, and one good way to make sure this happens in your systems is to provide short refresher courses.

EXERCISES

Develop a one-page specification for a testing/training database for typical users of a new text editing system that you are designing. Prepare a one-paragraph statement discussing how the testing/training database can be used to help identify and make changes to error-prone commands, procedures, messages, etc. of the new system.

8

HUMAN/COMPUTER INTERFACE FACTORS

INTRODUCTION

It is estimated that the human-computer interface factors generally contribute about the same number of errors as instructions or training factors. There is no doubt that their impact is large enough to be of concern to system designers.

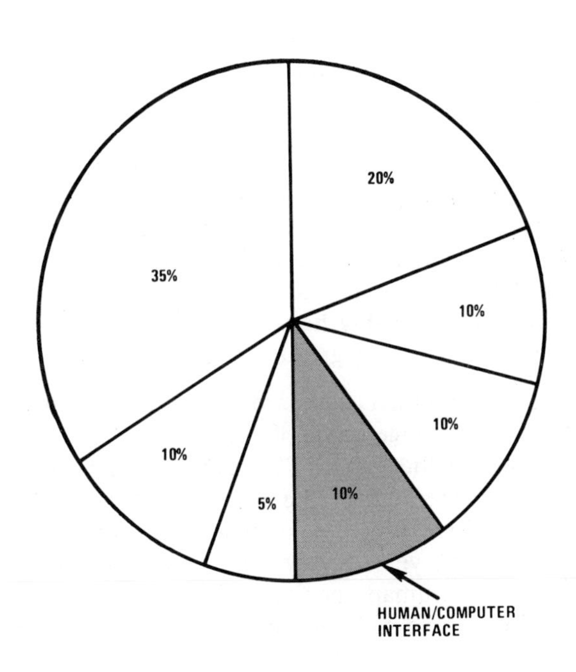

INPUT DEVICES

The equipment you select for system operators to communicate or interface with the computer could have a big effect on the number of errors made in your system. The most commonly used devices are video display terminals, which usually include a keyboard and CRT. You will probably never have a chance to design one of these interface devices, but you may certainly have many opportunities to select one from the hundreds that are available. Selection of human-computer interface devices must be a serious consideration of system designers, because an inadequate piece of equipment could certainly mean an increased number of errors.

For several years, human factors psychologists have been doing studies on many of there human/computer interface factors. They have made much progress in understanding what conditions and characteristics contribute most to having error-free performance. For example, we now know what should be the minimum size of printed characters, the optimal viewing distance for a CRT screen, and how high to place keyboards from the floor. Most input device manufacturers are aware of the need for having a good human/computer interface and have incorporated the results of human factors studies in their products. However, some producers of interface equipment still do not fully understand, and consequently do not address human performance considerations. For this reason, a system designer should carefully select the devices to be used.

A list of human factors requirements has been produced to help you select devices that will lead to the fewest errors. This list is shown in Appendix C. Probably the best way to use the list is to first narrow your equipment possibilities down to the four or five top contenders. After narrowing down the number of alternatives, evaluate each device in terms of each of the requirements shown. To help establish a priority of the requirements, some are obviously more important than others, a weighting scheme has also been provided. A "1" indicates requirements that will probably have the least effect on the occurrence of human errors, and a "5" the greatest effect.

Even the old familiar standard typewriter keyboard does not represent the best arrangement of keys. The standard typewriter keyboard is known as the QWERTY keyboard. The name QWERTY comes from the arrangement of the first six letters on the second row of the keyboard. The QWERTY keyboard arrangement has been the same for at least 100 years. Although it is probably not the best keyboard for good human performance, the standard QWERTY keyboard has been around so long that virtually everyone uses it.

The current front-running contender for a better keyboard arrangement is called the Dvorak Simplified Keyboard. The Dvorak keyboard, whose developer has been trying to have it adopted for at least 40 years, has recently been given greater attention. The Dvorak keyboard is arranged so that the letters typed most frequently are on the home row. Another consideration has been to lay out the keyboard so that as you are striking a letter on the right-hand side, your left hand can be searching for the next letter. According to its developer, when using the Dvorak keyboard, the fingertips of an average typist travel about one mile a day, whereas when using the standard QWERTY keyboard, the fingertips travel about ten miles per day. Nevertheless, the chances of all key operators changing from the QWERTY to the Dvorak keyboards is small mainly because too many people would have to go through a costly retraining process.

In order to ensure that system users and the computer effectively communicate with one another, it is important to design highly compatible human/computer interfaces. There are two "Interface" factors that are critical to your design. First of all, it is essential that much time and thought be given to providing good *message* and a good *command* language. At a minimum, this means that all messages must be optimized (so that the *computer can communicate with people);* and that the command language and data codes are also optimized (so that *people can communicate with the computer*).

MESSAGES

First, let's consider problems relating to the computer trying to communicate with people. Although it is certainly possible to build a computer that talks and even sings, in most data systems the computer still communicates by some type of visual display (i.e., usually a CRT). Thus, it is important that we take whatever steps are necessary to make the messages on a CRT screen as clear, concise and uncluttered as possible. None of us like to listen to people who ramble, and keep going on and on talking about things unrelated to our interests. And so it is with visual displays; they should provide only that information that people need. Some designers like to put extra information on the display "just in case". But good system designers find out in advance exactly what people need to know from the computer, and display only those items.

There is a special kind of CRT visual display that is generally called as "mask." A mask on a CRT screen looks much like a paper form that hasn't been filled in yet. In fact, that is the purpose of using a mask — to provide a way for system users to "fill in the blanks." Filling-in a mask is much like filling-out a paper form.

A good mask is one that enables a user to enter information quickly and with few (if any) errors. The same rules used for designing paper forms apply for designing CRT masks. It is very important to arrange the fields on a mask in exactly the same order the person needs to enter them. Frequently, a person sitting in front of a mask is reading from a paper form, and the paper form has data items arranged in a certain order. The mask should have the *same* order. Another major consideration is to have clear and meaningful headers for each field on your mask. Some system designers increase the number of errors in their system by trying to abbreviate headers on masks too much.

One final note needs to be made. The sequence that is most natural for a person to fill out a form may not be the most efficient sequence of information for the computer to process. A good designer will develop a form that optimizes and encourages a natural and error-free *human* interaction. At the same time, and quite independently, the designer should develop a set of computer processing routines that are efficient for the *computer.* Designing the software can be done without consideration of one another. Two different people may be involved – a human performance specialist may design the form, and a computer performance specialist may design the software. Once the form is optimized for the people who will be using it and the software is optimized for the computer, the space between the two can be easily bridged with a special interface program. To review, the most natural sequence for filling out a form may not be the best way for data to be efficiently processed by a computer. This apparent discrepancy can be dealt with by developing each separately, and having a special interface program bridge the gap between the two.

The other common way of receiving information from the computer is from paper printouts. You should pay particular attention to arranging the information on the printout in the *exact* order it will be used. We don't want a person having to read column one and then column ten and then back to column three. It used to be that some system designers would pay little attention to the needs of the user, and would literally "dump" all information in a database on to a paper printout, usually in the order it was arranged in the data base. There are even some printouts meant for system users that are coded in *machine language.* Either the system designer didn't know how to convert to English or else didn't care about human error in the system. A designer should always strive to optimize human-computer communication by making the messages and format of paper printouts and CRT screens as clear to the reader as possible.

COMMAND LANGUAGE

Another major human/computer interface consideration is the command *language* that is used. Obviously, if we used German, Chinese or Russian as the command language, we could count on having many errors until the system users became completely familiar with whatever language was used. The same is true if we use a machine language or modified machine language for day-to-day processing of information. The best possible language for English-speaking users is *English.* In most situations, you should use clear English commands.

Actually, the general rule is to use a vocabulary that is most natural for users of a system. For mathematicians, this may be a series of special mathematical notation symbols. For a chemist, it may mean sophisticated formulas.

DATA CODES

Even your input data, wherever possible, should be in English. Unfortunately, it is not yet always possible to have all our data items in English. Computer systems are using many (too many) non-English "codes." We are all familiar with the Morse code used by telegraph operators, and with secret codes used by spies during World War II (and even today). Codes are simply a set of symbols for communication. Some codes obviously help communication more than others. The two most common codes in data systems are command codes (usually abbreviated command words such as "sub" for "substitute"), and data codes, which are stored and manipulated by the computer.

Although it is not altogether a good thing, codes seem to be essential in computer-based data systems for identifying accounts, customers, products, employees, etc. They are useful because in general the shorter the identifier of a piece of data, the less costly the processing. It is obviously easier and usually less costly to continually process a person's employee number of 71432, rather than his name of Albert Quincy Haversackian. Less keying time and fewer key strokes are required by people, and less storage space is required in the computer.

Whereas codes facilitate the processing of data *in* a computer, many of these codes result in numerous errors that are difficult to prevent and detect when handled by people *outside* the computer. In general, codes in computer systems have some *good* attributes and some *bad* attributes.

Although many codes will be inherited by your system, you will probably have to design others. There has been a moderate amount of research on what a good code should look like. For our purposes, a good code is one that tends to elicit few errors. There is one code design rule that tends to be very important: keep codes *short.* In general, as the length of a code increases, so does the number of errors elicited by that code. An example from one study (MacArthur, 1965) is shown in the following table:

Number of Characters per Code	3	5	7	9	11
Percent of Error	1.4%	2.0%	2.6%	3.1%	3.6%

In this study when the average number of characters was five they had a field-level error rate of about two percent. But when subjects had to deal with 11 characters, their error rate was over 3.5 percent. The number of characters per code is definitely related to the number of errors that will occur in your system, i.e., longer codes mean more errors.

It is interesting how some system designers try to cram all kinds of "intelligence" into a single code. One wonders if anyone has considered how many errors are made because of having long codes that contain a little bit of information for "everybody." A good example is the New Jersey driver's license number. It is a code that is *fifteen* characters long. The New Jersey driver's license number, like many codes, is handprinted and re-handprinted numerous times by a variety of different people (an officer when issuing a citation or warning, court clerks while preparing a court docket and while collecting fines, banks and other places of business when trying to establish identification for cashing checks, etc.).

The question could be asked, why such a long, long number? I figured that the number of possible 15-character combinations is about 1,000,000,000,000,000. But New Jersey driver's license number also includes at least one character position for an alpha character or letter, which substantially increases the total number of possible 15-character combinations. When using all ten numeric characters in 14 character positions and all 26 alpha characters in one character position, the number of possible different numbers is 2,600,000,000,000,000.

New Jersey only has about four million licensed drivers, which means that there are about 2,599,999,996,000,000 numbers "left over". A code only half as long as 15 characters is all that is required to give each driver in New Jersey a unique code. In fact, every person in the *world* could be assigned his or her own number by using only a 10-character code. By using a 15-character code, individual numbers could be assigned to every man, woman and child on earth, plus every person on over *700,000 other worlds* with the same population as earth. The question could be asked, in view of the high error potential is it really necessary to use a 15-character code?

There are other rules that should be considered when designing codes. The most important of these are included in Appendix D and should be carefully and thoughtfully followed as new codes are needed. Before using the list of rules make sure you know *exactly* how the codes will be used by people in your system. For example, will they be handprinted, typed, or spoken over a phone? Will someone be expected to memorize them, or will they be used as simple "throughput"? After you know how they will be used, then the appropriate rules in Appendix D can be applied.

The factors associated with Design, Written Instructions, Training and Human/Computer Interface could account for over half (about 60%) of all the errors made in your new system. A system designer has almost total control over these factors. That is, decisions on these factors will either *haunt* or *help* user performance for the life of the system. The designer usually has the power and authority to make whatever decisions are appropriate in these areas, and system users must live with your decisions essentially "forever". Hopefully, these decisions will be well thought-out, and made only after much study and understanding. Keep in mind that poor decisions can lead to increased errors.

EXERCISES

Consider the following human/computer interaction. It involves an inexperienced user trying to figure out how to use a text editor:

USER:	start editor
COMPUTER:	?
USER:	edit
COMPUTER:	?
USER:	ed
COMPUTER:	?
USER:	help
COMPUTER:	Illegal Filename
USER:	what is filename
COMPUTER:	Illegal Command
USER:	what is legal
COMPUTER:	Illegal Command

USER: edit termpaper
COMPUTER: ?
USER: help
COMPUTER: Illegal Command
USER: ed termpaper
COMPUTER: ?
USER: help
COMPUTER: Illegal Command
USER: e termpaper
COMPUTER: _ (no response except a flashing cursor)
USER: e termpaper
COMPUTER: _ (flashing cursor)
USER: help
COMPUTER: _ (flashing cursor)
USER: stop
COMPUTER: _ (flashing cursor)
USER: quit
COMPUTER: _ (flashing cursor)
USER: (turns off terminal)

In this situation, the new user was frustrated by the computer to the point of giving up. At first, the computer would not accept the commands (start editor, edit, ed), and made only limited, awkward attempts to inform the user. Later, the user almost accidently entered the correct command (and legal filename), but again the computer did not inform the user.

Assume the same user tried to access a system that *you* designed. Using the same user inputs, what would be your (computer) responses? Show how the user would "access" the editor, "enter" text and finally "quit" the editor. Show the user's entries followed by your computer responses. Use the same format shown above, e.g., USER, then COMPUTER, then USER, etc.

9

ENVIRONMENTAL FACTORS

INTRODUCTION

We are now going to briefly discuss three factors that can have a considerable influence over the number of errors occurring in a system, but usually are *not* under the control of the system designer. These errors, for the most part, are under the control of the group that will be *using* the new system. It should not come as much of a surprise to learn that a designer can do an excellent job of optimizing conditions with the design, written instructions, training and human-computer interface factors, and there may still be a substantial number of errors in a new data system. The people using the new system must take over and continue making good human performance decisions. It appears that the responsibility for errors in a "typical" computer system must be about equally shared (50:50) by system designers and system users. Thus, if the system designer has done a good job of dealing with human performance issues, the majority of errors in a system probably can be attributed to poor decisions by system users.

System designers usually have the least impact on the following three factor categories: *environmental, organization accuracy requirements and personal factors.* Nevertheless, these three factors will be briefly discussed in order to complete our review of factors that are related to the occurrence of errors. It is highly appropriate and desirable for the system designer to discuss these areas with new system management. That is, a system designer has a much better chance of having a new system "look good" if time has been taken to ensure that the environment is adequate, the accuracy requirements are high, and that the right numbers and kinds of highly motivated people will be operating the system. So, indirectly the system designer can influence subsequent human performance decisions (those made after the system is delivered) by taking time to inform and discuss relevant items with the new system management.

THE ENVIRONMENT

Environmental factors are certain conditions that exist in the "world" or "life space" of a user, which affect the accuracy of his or her performance. These include fluctuations in room temperature, employees passing back and forth in front of a work station, or noise (any undesirable sound that tends to disrupt or annoy workers).

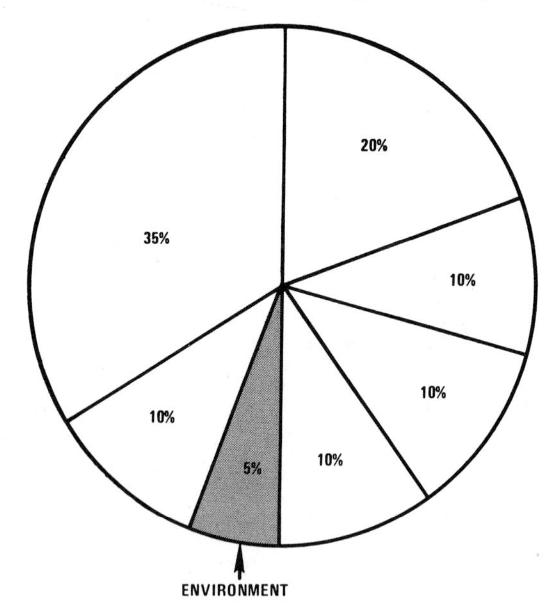

The environmental factor probably contributes less to the error level then any of the other factors. But it could contribute enough to warrant attention by users. For the most part, it is not under the control of the system designer. Nevertheless, if not optimized, extreme environmental conditions could lead to at least 5% of all errors in a computer system.

The environment in most systems consists of a "physical environment" and a "social environment". The main conditions in a physical environment that are of the most concern include room temperature, amount of light, and noise level. The main conditions in the social environment include being isolated or overcrowded, a user's relationship with other workers, a user's relationship with supervision, etc. The set of environmental factors of most interest includes all influences and conditions in the physical or social environment that could possibly cause errors to occur. Exactly how some of these factors affect performance is not very clear.

An environmental factor is here defined as any environmental influence or condition that increases the probability that errors will occur during the performance of a task. There seems to be an optimal environment in which people perform best.

PHYSICAL ENVIRONMENT

To help demonstrate the level of understanding and the complexity of the physical environment factors, a more detailed (although still very brief) account of research findings will be presented on the environmental factor of *noise.* A summary table is then provided showing other influences in the physical environment that may be related to increased errors. Noise has been selected for discussion because it is more likely to reduce the *accuracy* of work than to reduce the total quantity of work (Wilkinson, 1969; Miller, 1974).

Noise

When a task requires the use of speech, noise that masks or otherwise interferes with the perception of the speech will obviously degrade with the performance of a task. When tasks do *not* involve auditory signals, the effects of noise on their performance have been much more difficult to access. Nonetheless, certain general conclusions have emerged. One of the most interesting is that steady noises without special meaning do not seem to interfere with human performance unless the noise level (A-weighted) exceeds 90 dB to 100 dB, and performance on the task lasts for a least 30 minutes (Wilkinson, 1969). Thus, in order for noise to hinder human performance in a computer system, it has either got to be louder than 90 dB, or interfere with speech.

Most people do not know how loud 80, 90 or 100 decibel noise is. The following chart has been included to give you some idea of the relationship between decibel (dB) level and some common sounds:

Decibels	Common Sound
100	Subway train at 20 feet.
80	Passing truck
60	Conversational voice.
40	Quiet residence.
25	Broadcast studio.

The noise level at which human performance begins to be degraded is about the level of noise made by a subway train at twenty feet. As a result, not too many data system environments have noise levels that are of much concern.

A third consideration with noise is that irregular bursts are generally more disruptive than steady noises. Even when sound level of irregular bursts are below 90 dB, they may sometimes interfere with performance of a task. In fact, intermittent noise is usually much more distracting than continuous noise. One reason that has been suggested is that people can at least partially "adapt" to continuous noise if it is not too loud, but adaption does not appear to occur so readily with intermittent noise (Poulton, 1972). Another researcher has observed that a burst of noise that lasts for only one second may upset a person's performance for the next 15 to 30 seconds (Woodhead, 1959).

Three other considerations when dealing with noise are of interest (Miller, 1974):

a. High-frequency components of noise, from about 1000-2000 Hz, seem to produce more interference with performance than low frequency components of noise

b. Noise does not seem to influence the overall rate of work, but high levels of noise may increase the variability of the rate of work. There may be "noise pauses" followed by compensating increases in work rate.

c. Complex tasks are more likely to be adversely influenced by noise than are simple tasks.

Other Physical Environment Influences

Besides the effects of noise on human performance, there are a few other conditions in the physical environment of which a system designer should be aware. There seem to be certain limits in which the working environment should be maintained. These limits for several different conditions are shown in Table 9-1. When the limits are exceeded, there is likely to be a fall in efficiency, which in many cases may lead either directly or indirectly to an increased susceptibility to making errors.

SOCIAL ENVIRONMENT

In the social environment there are a couple of situations that deserve consideration. One has to do with over-crowding and isolation. Many people do not work well in over-crowded conditions, or in situations where they are absolutely alone. For other users, the number of people

Table 9-1. Environments Just Severe Enough to Produce a Reliable Deterioration in Performance (after Poulton, 1972)

Environmental Condition	*Condition just severe enough to degrade performance*
Heat	*Air temperature 27° C (80° F)*
Cold	*Air temperature 13° C(55° F)*
Dim light	*7 to 10 ft-c*
Glare	*Depends upon the angle of the glare source, of the line of sight, as well as upon its brightness*
Vibration	*0-.02 inch at 19 Hz*
Acceleration	*3G upward (gravitational force 3 times the normal size)*

in the work area does not really have much effect on their work. When trying to operate a computer system where either of these two extreme conditions exist, it may be well to find out in advance how potential users may feel about it.

A second situation of concern in the social environment has to do with the workers interaction with one another. It is interesting that at least one study has shown that people who make more errors are those that get along *best* with their co-workers (Bailey, 1975). It may be that a high level of co-worker inter-action is indicative of a "country club" atmosphere in the work environment. Some users gain the most satisfaction at work from associating with other people, whereas other workers derive more satisfaction from doing a "good job". It is not clear what all the social conditions are that affect human performance. The ones that have been briefly mentioned, over-crowding, isolation and co-worker interactions, represent only a few of many conditions and situations that actually do exist. But it does seem clear that the social environment in one way or another can have an impact on the number of errors occurring in a computer system.

FAR-OUT FACTORS

A word of caution. Don't waste your time trying to reduce errors in a computer system by trying to control some "far-out" environmental factors. Many lesser known environmental factors have received much study, and have been generally *discarded* as not having a great enough effect on human performance to take seriously. These include changes in barometric pressure, latitude and longitude, lunar (moon) phases, shape of a room, and the existence or non-existence of positive and negative ions.

EXERCISES

Develop a set of recommendations to improve the environment where the terminal you most often use is located. Consider both the physical and social environments. List at least ten changes that could help reduce errors, with the most important at the top of the list, the second most important second, etc. Be creative!

10

ORGANIZATIONAL ACCURACY REQUIREMENTS

INTRODUCTION

The organizational accuracy requirements, set either directly or indirectly by all organizations, have a relatively strong impact on the number of errors made in a data system. It is estimated that the strength of this factor is about equal to the training and documentation factors. For the most part, the organizational accuracy requirements factor is *not* under the control of the system designer.

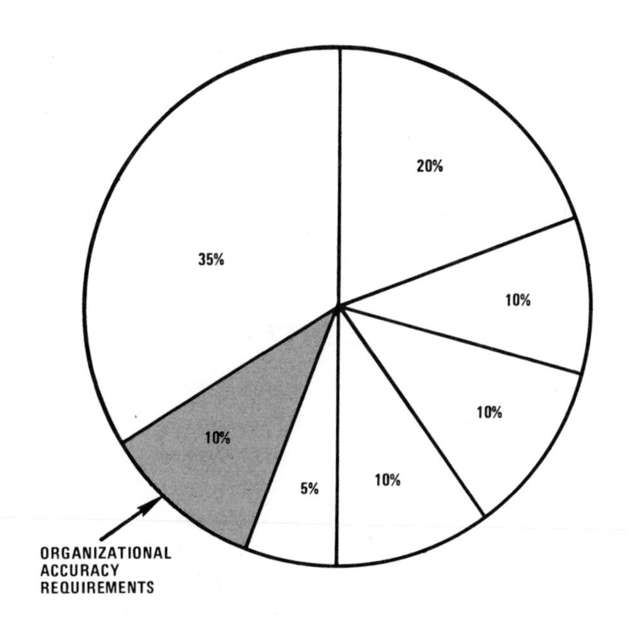

SETTING ACCURACY REQUIREMENTS

Every organization seems to have a different attitude about acceptable accuracy levels. Some pay little attention to errors, others take them very seriously. A few years ago there was a big move in many industries to have "zero defects" (Halpin, 1966), which translated to a company-wide objective of essentially *no errors.* If this position is established as an organizational requirement or policy by upper level management, and is actively enforced, the people operating data systems are more apt to strive for "zero error" performance. If management has taken a more relaxed view of errors, then people operating computer systems will also be more "relaxed", and more errors can be expected.

Many organizations have little or no idea how much errors are costing them. If they did know, it is possible that more emphasis would be placed on preventing their occurrence. There is no question that one effective means of reducing costs and preventing errors is to set high accuracy objectives, and expect these objectives to be met. If the objectives are taken seriously by management personnel (from top-level to first-level), the errors in systems will be greatly diminished. But keep in mind there are limits to the error reduction potential from this one factor. It is not the answer to *all* error problems.

The organizational requirements on data accuracy are usually communicated to workers in data systems primarily by the speech and behavior of supervision. It is always interesting to observe the "accuracy criteria" that has been set by supervision in a system. Some supervisors work very hard to keep errors down, while others spend more time doing other things. The first-level supervisor's behavior is usually a good reflection of the organizational requirements set (or assumed) by upper management. Their feelings about errors will become very apparent to workers by their actions (or lack of action). Also, supervisors have an opportunity to encourage the use of reference materials, conduct on-the-job coaching, watch for needed refresher training, etc.

PERSONAL ACCURACY CRITERIA

It is becoming more apparent that people set their own personal "accuracy criteria" for each activity in which they participate. Most people working in computer systems do not set this criteria at the zero-level. That is, users usually accept a certain proportion of their own errors as a normal and natural part of doing the work. Existing organizational policies or requirements have a strong influence on the level at which this personal accuracy level is set. If things are "loose" and errors are not taken too seriously by management, it is highly likely that they will not be taken too seriously by individuals working in the organization. In other words, management feelings about acceptable error levels tend to become employee feelings about acceptable error levels.

REPORTING

Management personnel can do a much better job of maintaining high accuracy levels if they have information from good management-level error reports. As they try to ensure that organizational accuracy requirements are being met, they should make much use of computer generated accuracy reports. A system designer should determine early in the design of a system all management needs for information on errors. There probably will be a need, for example, for reports on individual error rates, system level error rates, the number of unresolved errors waiting to be corrected, the average time required to correct each error, etc. One note of caution: if you ask management personnel what reports or computer-based information they *want* to monitor accuracy levels, you may get a much different answer than if you try to define their actual *needs.*

Your system will look best if the using organization sets high accuracy level requirements. Of course, this assumes that the system designer has done everything possible to ensure that human performance will at an acceptable level. If high accuracy requirements are set by the organization, and the system is *not* well designed there could be major problems. At first, the users of the system may be blamed, but it will not take long for management to realize that the computer system itself is faulty. It used to be that people were expected to operate efficiently with virtually any computer system. But it is now becoming more and more common for major, expensive and frequently embarrassing re-designs to be undertaken.

EXERCISES

The purpose of this exercise is to demonstrate the effect of having accuracy criteria. You will need four people to help with the exercise. Each person will use a typewriter or video display terminal (keyboard and CRT) for entering several codes.

Have each person enter both Codeset A and Codeset B; however, two people should enter Codeset A first and two should enter Codeset B first. When a person is ready to type, but before entering a codeset, read out loud the approriate instructions.

"Type the codes in Codeset A as quickly and as accurately as you can. Don't stop to correct any errors you may make. Ready? Begin."

Before a person enters Codeset B read these instructions:

"Type the codes in Codeset B *with total accuracy.* Make no errors. Ready? Begin."

CODESET A	*CODESET B*
S92BNCE846	48PJLS4C1D
YMO9PEV5SQ	INUPM8N9L7
P7M8B0SD3X	Q5GTXZ3C46
J9N7M8EB2A	ON7M9LMI2X
9M7NHPMOKJ	DE2X5ONU8M
KJIM82X4C5	2A4ZV0M8LY
INOP761QAZ	789PLMQRFG
SDF439M0N8	8UHKP24RQD
KO8N7WB61Z	14RCZUIO7M
R564WMN9PU	8YHNMCFR43

After all four people have typed both sets of codes, add up the total number of errors for each set. Did setting the strict accuracy objective seem to have any effect on the number of errors made? Write a one-page report describing your findings.

11

PERSONAL FACTORS

INTRODUCTION

The one factor that can contribute most to the number of errors in data systems is the personal factor. It's influence is probably greater than many of the other factors combined. For the most part, the actual people selected to perform in a computer system are *not* under control of system designers.

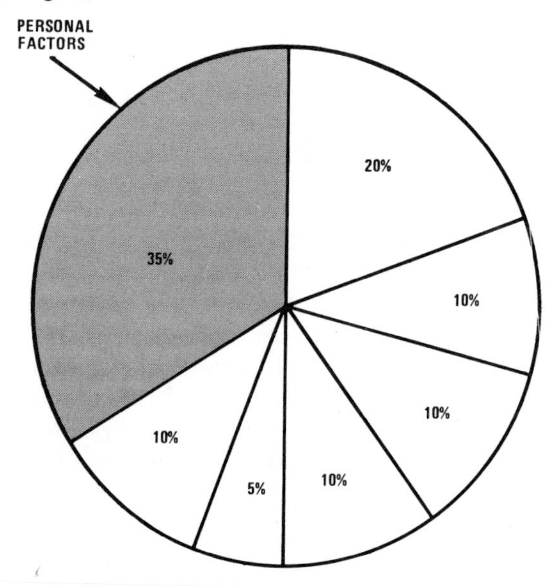

Some of the most complex and elusive causes of human error are contained in these conditions we are calling "personal factors": those

traits, characteristics or conditions peculiar to people that affect, in a relatively consistent manner, their ability to perform accurately. These include both physiological and psychological variables. These variables cannot be removed from the system unless the human component is replaced by a computer.

An erroneous assumption that is frequently made by system designers is that each and every person who will ever be involved in the operation of their new computer system will have perfect sensory, perceptual, memory, cognitive and movement capabilities. There is also an assumption of perfect physical and mental health. Designers also seem to assume that each person will have had adequate sleep and not be physically or psychologically fatigued. Finally, designers often assume that all system personnel will show up to work "wildly enthused" about spending an exciting day with the computer. None of these assumptions are *always* true. People come to work with a wide variety of ailments, imperfections and attitudes. And each worker in a computer system tends to be very much different (unlike the usually consistent and reliable computer).

MINIMUM SKILLS AND KNOWLEDGE

Probably the first and maybe even most important personal factor to consider has to do with the skills and knowledge of system personnel. Each user must have the minimum skill and knowledge required to benefit from training on the new system. If we misjudge our workers at this point, we will have problems for a long, long time. The workers may never learn what needs to be done, or how to do it. Thus, one of the most important "personal factor" decision, to be made in the operation of new systems is to select people that have the necessary minimum skills and knowledge.

A newspaper once printed the following ad: "WANTED - A person to work on a nuclear fissionable isotopic molecular reactor and three-phased cyclotronic photosynthesizer. *No Experience Necessary.*" Just finding appropriate people can sometimes be very difficult. And if those people selected for training do not have the necessary background or experience, the result could be (among other things) more errors. A system designer should keep in mind that it is very much possible to design a new system that requires such a complex set of skills that your organization could not afford to pay the people they would have to hire (if they could even find them).

As noted earlier, there will be tremendous individual differences among the people selected to work in new data systems. Some of these people will simply make more errors than others. One study of keypunch operators found that high error makers made almost *ten times* as many errors as those workers who made fewer errors (Klemmer and Lockhead, 1965). Obviously, if we can identify who the high error makers are, and keep them from working in computer systems, it will be to the advantage of all system designers.

STEREOTYPED BEHAVIOR

We have talked briefly about individual differences; there are also some individual similarities the affect human performance. Some of these are the user's stereotyped reactions. For example, a person filling out a form or blanks on a CRT in the United States expects to proceed from left to right and top to bottom. If the form or CRT does not allow this to happen, we have increased the probability of errors occurring. People get in the habit of doing things in a certain way and if the design of a new system forces them to do it in another way, errors will occur. For example, when turning on the light in a room, we usually expect to flip a switch "up". Flipping a switch "down" to turn on the lights seems unnatural. So it is with many tasks in computer systems. We should constantly try to find the *natural way* and design in that direction: To match the way an activity is to be performed with the expectations of users.

PHYSIOLOGICAL AND PSYCHOLOGICAL NEEDS

Another thing we have to consider when talking about people working in systems are their physiological needs. Unlike computers, people get hungry and frequently take breaks in the morning, afternoon and for lunch. They get thirsty from time to time and must leave their work to get something to drink. They stop work to use the restroom a few times each day. These are all very real considerations, because every time their work is disrupted, people have to return and become reoriented (warmed-up) to the work they were doing. This continual reorientation increases the probability of errors.

People also have a wide range of psychological needs. For example, people require approval and need to have feelings of self-worth. If a person is getting a lot of criticism from a supervisor or is not being accepted by other employees, it may show up in the quality of work.

SENSORY PROCESSES

Another set of possibly degrading conditions, that are frequently overlooked, has to do with sensory processes. Not all people see or hear at the same level. If we have a situation that requires people to do much reading, and if one or more of the people cannot see very well, many errors could occur. Or if one or more users are hard of hearing, particularly if they are not aware of this condition, it could also lead to many errors when collecting information from customers. All system designers hope and expect that if a person needs glasses for reading, and shows up to work one day without the glasses, they would not allow the person to work until the glasses are secured. Unfortunately, system designers have little control over such situations.

SLEEP LOSS

An influence that has received much study is the effect of sleep loss on human performance. It is interesting that adequate performance on most tasks (accuracy and speed) can be maintained for a short period of time in the face of quite severe loss of sleep. But as soon as the duration of a repetitive task is extended to 30 minutes or more, it becomes very difficult (if not impossible for some people) to sustain concentration — even when only one night of sleep has been lost (Wilkinson, 1965). Another investigator found that when the testing period was a full working day, efficiency on calculation and vigilance tasks was significantly reduced when subjects had two hours sleep on a single night, or five hours or less on two successive nights (Wilkinson, 1968). Again, we have a personal factor over which system designers have essentially no control.

It is interesting that sleep loss also leads to other types of problems. For example, when a person has lost sleep, the resulting drowsiness may contain lapses lasting about one second, during which the person is totally out of touch with the surroundings. In other situations, people who have dropped off to sleep and then were suddenly awakened take from one minute (Webb and Agnew, 1964) to five minutes (Langdon and Hartman, 1961) for reaction time to return to normal. Also, it usually takes at least one week, and sometimes longer, for a person to totally adapt to work nights shifts versus working day.

DRUGS

People use a variety of prescribed and non-prescribed drugs in their daily lives. These range from the common and extensive use of tea, coffee and tobacco, to the much less common use of heroin and morphine. Drugs that improve performance are generally considered as "stimulants", while those that degrade performance are considered "depressants." Although it is not yet clear, it is possible that common stimulants such as caffeine (coffee and tea) and nicotine (tobacco) reduce errors, at least while their effect is stimulating. But as their effects wear-off (between doses), *more* errors occur than if the stimulant had never been used in the first place. Most of the commonly prescribed behavior-related drugs are depressants. These include most tranquilizers, pain killers and sleeping tablets. Depressants tend to slow down responses more than cause errors to occur, but the true extent of their effect on errors is still not clear.

In some situations the effect of a drug may not impair performance until after its primary effects have begun to wear off. In one study on the after-effects of taking two sleeping tablets (McKenzie and Elliot, 1965), one group of subjects took the tablets (200 milligrams of secobarbital), slept for their regular time, and then performed a task throughout a regular working day. Another group of people who did not use the sleeping pills did significantly better, not just in the beginning, but for the *entire seven-hour day.* At the present time, it is almost impossible to tell when drug effects begin and end.

For the vast majority of available drugs, there is little or no research reported concerning their human performance effects. This same observation also holds for another subset of items that are becoming recognized as "occupational hazards" in some industries. These include materials that could be ingested, inhaled or assimilated through the skin such as solvents, pesticides, metals, toxic gases, etc. Any one or a combination of these agents could potentially degrade performance and lead to increased errors.

ILLNESS

Research on the behavioral effects of illness (with infectious diseases) is a fairly new area of investigation. As a matter of historical interest, however, it should be noted that as far back as 1937 a study was conducted using people who had tuberculosis. They found that TB victims had a tendency toward an increase in errors. More recent researchers have reported that decrements in task performance first appear in coincidence with the onset of an illness and become maximal when symptoms are greatest (Morgan and Alliusi, 1972). However, some people showed no decrements in performance whereas others were unable to work at all during peak illness. They concluded that human performance in a work situation seems to vary as a function of the *subjective symptoms* rather than the *objective indicators* of illness. It is frequently overlooked that substantially more errors are probably made by sick people who struggle to make it to work feeling that they are actually doing the company a favor.

BODY RHYTHMS

Other personal factors are biological and circadian rhythms (Colquhoun, 1971, 1972), including the menstrual cycle (Sommer, 1972; Wong and Tong, 1974). Many researchers have demonstrated that people tend to make more errors in different phases of these cycles. These studies suggest that there is possibly greater susceptibility to internal or external stressors at different times of the day or month.

FATIGUE

We should carefully evaluate the effects of fatigue on system personnel, taking steps to provide for regular rest periods, especially from routine monotonous work assignments. In addition, extended overtime assignments should be avoided. Fatigue seems to lower the tolerance of people and to leads to increased errors.

MOTIVATION

One of the most critical personal factors relates to motivation. Specifically, how does a person feel at the very moment a job is being done. The workers may be apathetic, completely satisfied, or enthused about what they are doing. As noted before, one of the biggest problems we have is that many system designers think that all workers will be exhilarated when the new computer system arrives and from that point on will be greatly enthused about their jobs. The fact is that many workers are probably more interested in things other than their work – even if it includes a computer. Even when they are enthused, the enthusiasm does not usually sustain itself over a long period of time. Thus, it is probably a good idea to develop a system that can be easily operated by people who are only moderately motivated.

There are, however, some things a system designer can do to help increase worker motivation. Motivation seems to be "helped along" when (a) the workers feel responsible for maintaining some level of efficiency, (b) each person is aware of the ultimate necessity for accuracy, and (c) definite accuracy objectives have been set, preferably somewhat above each worker's typical performance level. There are many books written on worker motivation, and much new research is being reported. Even so, few specific rules for increasing motivation are available to computer system designers.

Nevertheless, the way a system is designed can either release or inhibit the motivation to work. In fact, to a large extent worker motivation depends more on the content of the work (i.e., the design of the system) than on other conditions that surround it. In general, work that is either too easy or too difficult will tend to be less motivating than work that is slightly challenging. Some data systems are designed with work that is so simple or so difficult that most people dread to do it.

Designers should try to develop ways of getting the work done in ways where jobs are "enriched" as much as possible. Essentially this means to develop job content so that individuals have increased responsibility and autonomy, a wider variety of tasks, and more opportunities for achievement and recognition. In effect, the *work itself* becomes a source of motivation for the individual (Ford, 1969, 1973; Hertzberg, 1968).

COMBINING PERSONAL FACTORS

Earlier we talked about the effects of combining two or more "out of tolerance" environmental conditions. When combining two or more personal factor conditions, the results are just as interesting. For example, when combining "loss of sleep" with a moderate dose of a drug like alcohol, human performance is *impaired.* With people who have had normal sleep, the same dose of alcohol has *no effect* (Wilkinson, 1969). Similarly, when "reduced incentive" is combined with "loss of sleep" human performance is *reduced.* But when "loss of sleep" is combined with "increased incentive" it has *no effect* on performance. Some factors seem to balance each other out. When "incentive" is combined with loud noise, people tended to make more errors with *increased* incentive than with *decreased* incentive. The outcome of two or more interacting conditions is not easy to predict.

CONCLUSION

The system designer usually has little impact on the environment, organizational accuracy requirements and personal factors. Nevertheless, these three factors have been briefly discussed in order to complete our review of factors that are related to the occurrence of errors. It is highly appropriate and desirable for the system designer to discuss these areas with the new system management. The system design has a much better chance of "looking good" if a system designer takes the time to ensure that the environment is adequate, the accuracy requirements are high, and the right numbers and kinds of highly motivated people are operating the system.

EXERCISES

Determine the population stereotypes (i.e., preferences) for the following command structures. These are commands that are commonly used with text editor systems. Ask at least ten people which command structure they prefer for each action, option A or option B. Select five people who have experience with computers, and the other five that have no computer experience. Keep a record of their responses, and after everyone has given an opinion summarize your results. Prepare a one-page report of your findings. Is there a strong population stereotype in favor of verb-object or object-verb? Is the stereotype different for experienced vs. inexperienced people?

DESIRED ACTION	*COMMAND OPTIONS*
To remove a word	A- remove word B- word remove
To add a paragraph	A- add paragraph B- paragraph add
To substitute one word for another	A- substitute oldword newword B- oldword newword substitute
To delete a word	A- delete word B- word delete
To replace a word	A- replace word1 (by) word2 B- word1 (by) word2 replace
To insert a sentence	A- insert sentence B- sentence insert
To change a word	A- change word B- word change
To remove a paragraph	A- remove paragraph B- paragraph remove
To find a word	A- find word B- word find
To delete a file	A- delete file B- file delete
To run a program	A- run program B- program run
To enter a document	A- enter document B- document enter

12

ERROR DETECTION

INTRODUCTION

In most computer-based systems, errors can be detected either by people or the computer. A good system designer will use both means of error detection in order to catch as many errors as possible. Obviously, the less attention given to error *prevention* the more attention a designer will have to give to error *detection*. But even where many errors are prevented from ever occurring, the designer should seek to detect those few errors that are still being made, or that may be inherited from other systems.

There is no question that the computer, if properly programmed, can detect certain types of errors. But when trying to control errors in a new computer system, it is a mistake to emphasize error *detection* over error *prevention*. Some system designers have failed to control errors because they have tried to overwhelm complex error problems with simple software solutions.

People self-detect many of their own errors, while other errors can be detected by proofreaders. However, most people have difficulty detecting errors made by others. If we were to rely only on manual proofreading to catch the unpreventable errors in computer systems, our systems would contain a substantially greater number of errors. Fortunately, numerous computer-based detection techniques have been devised over the years and many more will be developed. The computer has greatly increased the capacity for detecting human errors. However, even the most sophisticated and elaborate computer-based error detection techniques *cannot* catch all the different types of errors that people make. In fact, one of the most common costly mistakes made by many system designers is to assume that the computer will catch almost all human errors.

This chapter is intended to review some ways of detecting errors in computer systems. A system designer should be aware of these techniques, then, considering the system requirements and cost tradeoffs, develop a workable plan for detecting as many errors as possible.

Several error-detection techniques will be discussed. By reviewing these techniques, it should provide you with ideas of a variety of different ways to catch certain types of errors. The techniques could be used "as is" where appropriate. However, many of these techniques could be expanded to become even better ways of detecting errors. In this sense the concepts to be discussed may be more important then the techniques.

FINDING ERRORS FAST

Errors should be detected as soon as possible after they are made. Among other things, this means a system should be designed to facilitate *self-detection* of errors. If an error is not self-detected, then it should be detected as quickly as possible thereafter by a computer. Many "intelligent" video display terminals help to quickly detect certain kinds of errors. Many of these terminals are able to perform some simple edits and validates, and after detecting errors, give almost immediate error feedback to terminal users. So the fastest means of detecting errors is self-detection, and a close runner-up is detecting errors using an intelligent terminal.

Once data enter the computer even more edits and validates can take place, and if the system is on-line, the error-maker can know about certain errors soon after they are made. If, however, the computer processes in a batch mode, people will not find what errors they have made until later (e.g. an hour or day or longer). Finally, if an error is detected by a customer, or someone outside the system boundaries, the error-maker may *never* know that the error occurred.

There are at least two advantages to catching errors soon after they are made. First, if an error is detected soon after being made, there is a better chance of changing the error-producing behavior of the user who produced it. The longer it takes to detect an error, the less chance we have to change the errormaker. Second, if detected soon after being made, the correct response or right answer is more likely to be known. Generally, the longer it takes to detect an error, the harder it is to correct.

EFFECTIVENESS OF COMPUTER DETECTIONS

Let's assume that you set out to design a series of computer-based error detection routines, beginning with simple edits and proceeding to some limited, but rather sophisticated, table checking and data comparisons. Even with the best current technology and ideas, usually about half of the errors that come to the system are detectable. It is fair to say that most commonly used computer-based error-detection schemes detect far less than half the errors that enter the computer. The main reason is that computers are not very good at detecting the kinds of errors that people make most frequently.

COST TRADEOFFS

Even though the overall success at detecting errors with a computer probably will be 50% or less, there may be some critical data fields where a much higher percentage is required to be detected. For example, we may need a person's social security number to be accurate virtually *all* the time in order for a system to function properly.

In this case a series of elaborate and involved human and computer-based edit and validation routines will have to be developed. Detecting each error will be costly. As the errors become fewer they also become more difficult (and expensive) to catch. Some of these errors actually could cost far more to detect than they are worth. There always may be some errors that defy detection, and must be caught by a customer, or perhaps never be detected at all. Obviously, it is more cost effective to try to prevent these errors from ever occurring. Even so, the vast majority of errors in critical data fields usually can be detected if a designer is knowledgeable, creative, and has essentially *unlimited financial resources.* The system designer will have to decide when it is no longer cost effective to detect those last few errors.

The higher the proportion of errors that must be computer-detected the more time consuming it will be to design, and the more expensive the system will be to operate. A good set of accuracy objectives will help determine when to stop trying to detect more errors. If possible, there should be an accuracy objective (e.g., 99.8%)

for each data field. If the accuracy objectives are high they will be costly to achieve. Thus, the *benefits* from having fewer errors must be compared or traded-off with *total cost* of reaching a certain accuracy level.

DETERMINING DETECTION METHODS

Probably the best way to determine the number and kinds of error detection methods a new system will need is to do a small study. This study should be a survey of the number and types of errors that people are making in the present system. Consider, for example, a study done by the telephone company for this purpose.

They were developing a new computer-based system for constructing telephone directories. As part of creating a new directory each year, they were interested in detecting and correcting as many errors as possible. To do this properly it was important to know what errors were being made when people (a) requested a new telephone, (b) asked to change a name, address or telephone number on an existing telephone, or (c) requested their telephone to be disconnected.

The objectives of the study were to:

1. Identify the number and types of errors being made in the existing system
2. Determine the best means for detecting each type of error
3. Identify the number and types of errors that *cannot* be detected
4. Propose efficient ways of correcting all the different types of errors that occurred

Most of the errors in this new system were made by someone outside the system boundaries. This provided little opportunity to *prevent* errors from occurring. However, this was an "error-end" system, meaning that as many errors as possible needed to be detected and corrected. All the errors made and manually detected in a two-month period were evaluated. The researchers recorded both the correct and incorrect form of each error. They also determined the relative frequency of each type of error. An example of the error study results is shown on the next page.

Some of these error types could be relatively easily detected by the computer, others required manual detection (if they were detected at all). But the designer knew *in advance* the number and types of errors to plan for, and could make the best possible decision for each.

TYPE OF ERROR	EXAMPLE	ERRORS PER 10,000 SERVICE ORDERS
NAMES REVERSED	*C: HARRIS, GLADYS E: GLADYS, HARRIS	2
NAME OMITTED		2
LAST NAME INCORRECT	C: PHAM, KIM E: TRAN, KIM DUNG	2
	C: McGREW, THOMAS E E: M*cC*REW, THOMAS E	25
	C: GARITN, TERENCE E: GART*N*, TERENCE	43
HOUSE NUMBER	C: 804 N BAMBY AVE E: N BAMBY AVE	3
STREET NAME PREFIX	C: 85 W MILLER ST E: 85 *F* MILLER ST	15
NAME OF STREET	C: 3202 CALLOWAY DR E: 3202 CALL*W*OWAY DR	15
	C: 3410 S WILTS CIR E: 3410 S WILTS *CR*	14
TELEPHONE NUMBERS	C: 871-8174 E: 8*6*1-8174	14
	C: 277-4412 E: 277-*412*	4

* C= CORRECT, E= ERROR

MANUAL DETECTION

Self-Detection

Once an error is made, the most efficient means of detecting the error is through *self-detection.* Self-detection is not only the first line of defense, it is frequently the *best.* When keying input, for example, experienced users can self-detect up to 80% of their errors. Everything possible should be done to optimize conditions so that more errors are self-detected. If the proportion of self-detections is increased, the user is even more aware of the number and types of errors made. Self-detection also leads to more immediate error correction, makes the correction easier, and prevents an error from contaminating other data. There is no question that the most potent of all error detection techniques is self-detection.

People self-detect errors best if certain conditions are optimized. Some of the conditions to be considered include *appropriate feedback*, (ensuring that people can see or hear what they enter), *experience level* (people can build a self-detection skill if given a chance to practice), and *time constraints* (users must have time to make the detections). All of these conditions are under the control of the system designer.

Any effort to optimize conditions to facilitate self-detection will carry with it the payoff of reduced errors. Although it is not always recognized, people quickly and efficiently self-detect many errors that would require substantial computer processing for the same result. When compared with most computer capabilities for error detection, under optimal conditions people are far superior in detecting their own errors.

As noted earlier, errors made while keying words are easier to self-detect than errors made when entering meaningless data. With extensive use of meaningless codes, it is difficult to self-detect many types of errors. For example, a user could key or handprint the following list of last names and be able to self-detect many errors:

Should Be	*Entered As*	*Error?*
JONES	JOENS	Yes
SMITH	SMLTH	Yes
MASON	MASN	Yes

But if the same names had been previously converted to codes to facilitate computer processing, there may be the following situation:

Should Be	*Entered As*	*Error?*
893214	895214	?
HZ27917	HZ21917	?
237642	236742	?

There is no question that in order to facilitate self-detection (and effective proofreading) it is desirable to have codes that are meaningful to experienced users.

Standard clerical practices and well designed forms impose procedural controls on the creation of data. For example, assume a part number such as AX32621 is to be written on a document. Seven boxes may be printed on the forms. This is the exact number of spaces required for the part number to be entered. Any clerk writing a part number containing more or less characters than the required seven should notice the error. A well designed form will increase the proportion of certain types of self-detected errors, whereas a poorly designed form will make it very difficult to self-detect errors.

People seem to be motivated to self-detect some errors much better than others. Consider the employee who found that his paycheck was $5 *short.* He immediately went to discuss this problem with the people in the payroll department. They checked his records and found that the week before the employee had been *over-paid* $5. When asked why he had not pointed out the past weeks overpayment, the employee stated that he could tolerate an error once in awhile, but not when it became a habit. As this story suggests, people are biased error detectors. People are particularly good at "detecting" and correcting errors that benefit them. A designer should try hard to create a system where users benefit from catching their own errors.

Proofreading

When the telephone company puts out its Yellow Pages, the company tries to make sure it is totally accurate. Much of the checking for accuracy is done by proofreaders. Unfortunately, a recent edition of the Yellow Pages in a western city listed the Democratic Headquarters under the heading of "Amusement Companies." This error was not detected by any of the proofreaders.

For the most part, people are not very good at detecting other's errors. For a variety of reasons, some errors almost defy manual detection. It is also true that the fewer the number of errors to be detected, the harder it is to detect those remaining errors. Many years ago, Edgar Rice Burroughs, the creator of Tarzan, wrote a science fiction book entitled, "The Pirates of Venus" (Burroughs, 1934). One character in the book came to the painful realization that he had made a big mistake, possibly a lethal mistake, because he (like so many proofreaders in systems) had somehow "overlooked the obvious." Upon this realization he stated:

> To my mind flashed the printer's story of the first perfect book. It had been said that no book had ever before been published containing *not a single error.* A great publishing house undertook to publish such a book. The galley proofs were read and reread by a dozen different experts, the page proofs received the same careful scrutiny. At last the masterpiece was ready for the press — errorless! It was printed and bound and sent out to the public, and then it was discovered that the title had been misspelled on the title page (p.23).

If you are not yet convinced that people tend to have difficulty in proofreading to detect errors, consider the following exercise.

Study carefully this statement:

FEELING FINE AT THE END
OF THE DAY IS EASY WHEN
YOU ARE THE SON OF THE
FATHER OF FRANKENSTEIN.

If you were a proofreader and asked to find all the F's, how many would you say there are?

You should have found seven. Most people find only four. It is easier to identify errors in this exercise than when reviewing, for example, a list of insurance policy numbers because it contains words with meaning. It is far more difficult to manually detect an error in a code containing numbers and letters combined in some apparently meaningless fashion.

Another way of manually detecting errors is to have experts review computer printouts. One good example of an "error-through" system that really should be an "error-end" system is one that collects and stores personal data on people. Actually very little is done to check the accuracy of information inherited by these systems. This is particularly true of systems that maintain credit files on people. One study indicated that firms who maintain "people files" had a "surprising error problem" (New Scientist, 1973). About 75% of the companies surveyed admitted to finding serious errors when converting to a computer-based system, and 38% admitted to "significant problems in maintaining accuracy of these computerized records."

If they were true "error-end" systems, a major effort would have been made to detect and correct *all* errors. One obvious way to help maintain a high level of accuracy would have been to have employees check (proofread) their own records. In the same study of 43 organizations, it was interesting to note that in 9 organizations the workers did not even know they had a record, in 14 organizations they knew they had a record but were not allowed to see it, and in 4 organizations they could see only part of it. In only 15 (37%) cases had the system been designed so an individual could see and validate his or her entire file.

People receiving system outputs are important error detection control points, and provisions should be made for receiving feedback from all output recipients. Users of output may be people within the department, from other departments, or even outside the company (customers). Many White Page telephone directories, for example, are produced by computer-based data systems. In these cases, most users are customers that are not only outside the White Page Directory Department, but also outside the telephone company. Customers who find errors in their directory listing are encouraged to point these out to the telephone company. The errors are then corrected in the next issue. If users of your system outputs are not provided with convenient ways to feedback errors, certain errors could remain indefinitely.

Proofreading in one form or another is usually a part of every system. Designers should provide material to be proofread in a form where errors can be easily detected. The proofreading process can extend all the way to where information is used, as long as a means is provided to have all detected errors corrected.

COMPUTER DETECTION

Check Digits

One common way of using the computer to detect certain types of errors is with *check digits.* The check digit concept can be used with codes that only contain numbers, such as credit card account numbers. There are many different ways to calculate a check digit. We will consider one approach called "Modulo-10."

Assume that you have a seven digit account number and you wish to develop a check digit for these numbers. The check digit is generated for each code by multiplying the first digit by 7, the second by 6, and so on until the last digit is multiplied by 1. A system designer then uses the "Modulo-10" complement of the answer as the check digit and places it after the seventh number. The original seven character number now has eight characters, with the last character being the check digit.

For example, for the account number 1234567, the check digit would be calculated as follows:

Account Number: 1 2 3 4 5 6 7
Check Digit Multipliers: 7 6 5 4 3 2 1
Check Digit Computation:

1 x 7 = 7
2 x 6 = 12
3 x 5 = 15
4 x 4 = 16
5 x 3 = 15
6 x 2 = 12
7 x 1 = 7
Total = 84
Final Digit = 4
10 Complement (10 − 4) = 6

Thus, the number assigned as a check digit would be 6, and the new account number would be 12345676.

The Modulo-10 check digit technique was originally designed to catch single-digit errors (e.g., should be 12345, but entered as 12*6*45), and character transpositions (e.g., should be 12345, but entered as 12435). However, after many years, it was realized that there was one type of transposition that was not being detected (Taylor, 1975). Consider the number 12345676 that was just calculated. Remember that the final "6" was the check digit. If the last two characters were transposed when being entered, e.g., 12345667, it would still be a valid number. In fact, it turns out to be a valid number complete with its own check digit.

You can see why it is important to know the weaknesses as well as the strengths of any computer-detection method chosen for use. Another weakness of the Modulo-10 technique is the use of 5 as a multiplier. Any *odd* number multiplied by 5 always has an answer that ends in 5, and any *even* number multiplied by 5 always ends in 0. Errors made in the digit position where 5 is the multiplier will only be detected when an odd number is substituted for an even number, or vice versa. This could mean that about 50% of the errors made in the digit position where 5 is the multiplier will not be caught.

A check digit technique that supposedly overcomes some of the weaknesses of the Modulo-10 approach has been proposed (Taylor, 1975). This technique uses 2 as the multiplier for the first, third, fifth, etc., digits, and a 1 for the second, fourth, sixth, etc. digits. The other major difference is that the products obtained from multiplying by 2 are also added.

Consider the following example:

Account Number: 1 2 3 4 5 6 7
Check digit multipliers: 2 1 2 1 2 1 2
Check digit computation:

1 x 2 = 2		2
2 x 1 = 2		2
3 x 2 = 6		6
4 x 1 = 4		4
5 x 2 =10	1 + 0 =	1
6 x 1 = 6		6
7 x 2 =14	1 + 14 =	5
		26

Total = 26
Final Digit = 6
10 Complement (10 – 6) = 4

If a check digit technique is used for one or more codes in a computer system, it should be done with caution. A designer should be alert to the types of character errors that can and cannot be detected. Don't oversell the error detection potential of this technique. Remember that check digits are most useful in detecting the types of errors that people make least frequently (transcription and transform errors). One final word of caution. By adding a check digit, the code is made longer. Every extra character that people have to enter increases the probability of an error being made on that code. In some instances, people may make more errors because of having to enter the check digit then the check digit will even detect.

The check digit technique, like most computer-detection methods, can be used to detect both unintentional and *intentional* "errors." Obviously, the major concern is to locate unintentional errors, but there is a growing need in some systems to detect intentional errors, either fraudulent or malicious.

Evaluating a code's check digit can be done by an intelligent terminal, or it can be done in a host computer. The use of check digits with input devices has the advantage that incorrect codes are detected earlier when they can be more easily corrected. Also this provides more immediate feedback to the user. In addition, fields checked by check digits generally do not need to be verified after being keyed. Check digits are widely used with account numbers, employee pay numbers, and bank account numbers.

Redundancy

One of the most used and effective ways of ensuring reliability in computer systems is to build-in *redundancy*. The term redundancy is defined as "an act or instance of needless repetition." But in systems where accuracy is of great concern, redundancy is not always considered "needless." Redundancy can be used to help increase the reliability of many systems. The effective use of redundancy means that the exact same thing is not done just once, but is repeated two or more times.

Probably the most well known example of using redundancy in computer systems is the key and verify (verification) process. In this process, one person keys in a set of data, and a second person keys in the *exact same data*. The entries are automatically compared, and where two entries do not match, one of the two key operators decides which entry is right.

Redundant systems are quite common. Most commercial and military aircraft, for example, have many redundant systems. A use of redundancy in aircraft is the use of two or more different fuel tanks. If one starts to leak, there is always a backup. In the United States space program involving astronauts, two or more computers are used to perform exactly the same functions. One computer serves as the primary computer and is used until it malfunctions, at which time a second or even third computer is used. The idea of redundancy is also common in natural systems. For example, Gilb and Weinberg (1977) have rightly pointed out that

> For a bird, finding a mate is too important to be left to a single performance of a song. To a flower, propagation of the species is too important to be left to a single seed. To our bodies, the transmission of nervous messages is too important to be left to a single cell (p. 113).

When using the concept of redundancy to detect system errors we should consider both simple and complicated uses. Occasionally, it may be worthwhile to have a user enter the exact data twice, once right after the other. For example, the amount a customer pays on a bill could be keyed twice by the *same* operator (e.g., 27.32 space 27.32), and the computer could match the two. If the two numbers match, the amount is accepted; if not, an error message is transmitted to the error maker.

By its very nature, a redundant process involves duplication or even triplication. It may, therefore, double or triple the cost of data entry. Because of the cost, a system designer should carefully consider alternatives to verifying all data items.

One way of reducing the need to verify all data is to verify only if the keying error rate is above an acceptable level. Each operator's work is checked on a sample basis. If the error rate is acceptable, no verification is made of future work; if not acceptable, there is complete verification.

Another way to reduce data entry costs is to be selective in deciding what and when to verify. Generally, not all fields require the expense of being entered two or more times. A system designer should concentrate any "redundancy efforts" on *critical data fields,* those fields that will hurt overall system performance most if they are in error. In other words, we should seek to identify those data fields that are most critical to the success of the system, and then decide if extra error-detection effort is truly required and would be cost beneficial to the system. In cases where near perfect accuracy is desirable, the use of redundancy should be seriously considered.

Of course, redundancy does not only mean that an item will be repeated twice. There may be super-critical data fields where a person may be required to enter the same data three or more times to help ensure accuracy. With this type of redundancy, like all forms of redundancy, the repeated information may or may not take the same form. For example, when we make out checks we write the amount in two different ways: $10.25 and ten dollars and 25/100 cents. A super-critical expiration date may be entered as November 9, 11-9 and the Julian date 313. All three can be compared by the computer and must match, or an error message is presented.

We can build a strong error detection capability by using several different combinations of edits and validates. If we use only the techniques concepts of check digits redundancy, I should be able to effectively eliminate certain types of errors. Unfortunately, we are generally dealing only with errors made at data entry (i.e., at the keyboard). And, as you may recall, less than 20% of all the errors that could possibly reside in your system probably occur at this point.

Source Redundancy

In addition to keying and verifying, the concept of redundancy can be used in other ways to help reduce errors in computer systems. Keep in mind that the double entry of certain *critical* data items also may be made by users who are not sitting at a terminal of some kind. For example, this could include people in your system who are talking with customers and filling out forms. These users could be at the very edge of the system boundaries.

The technique of redundancy probably leads to the most detected errors if used as close to the true source of the data as possible. As noted earlier, a common example of writing the same data twice is found when writing a check to be cashed at a bank. In this case, we are actually *outside* the bank's system boundary, but the system designer was wise enough to force us to comply with certain rules when filling out the form (check). As designers we should constantly be "shoving" the responsibility for errors back to the potential error makers. It would be foolish only to use redundancy techniques at the input terminal. Remember that when compared with all sources of error, only a small portion of all errors occur during actual data entry.

Gilb and Weinberg (1977) have presented a good illustration of using redundancy at the source:

> If a certain batch contains 412 chairs, we would like that number, 412, to reach the computer. Suppose an inventory clerk counts 409 chairs and writes this count in two places on a "source" document. Will this duplicate writing protect against errors?
>
> It will protect against errors that occur *after* the counting, such as:
>
> (a) the clerk miswriting one of the 409's, or
>
> (b) a data entry person misreading 409 as 407
>
> In no case, however, can the repetition help the computer obtain the correct number of chairs.
>
> To eliminate counting errors, the duplication must be *earlier than the counting,* or *in the counting itself,* by some such means as:
>
> (a) having two different clerks count the chairs,
>
> (b) counting the chairs once and deriving another count from a tally of parts going into the chairs, or
>
> (c) having one clerk count twice, but by a different method each time

> These forms of repetition would protect against the above errors, plus errors in the counting process itself. (p 121)

Consistent with this illustration, a system designer should add redundancy as close to the source of new information as possible. Another good way to do this is by building-in "checkwords".

Checkwords

The concept of a "checkword" was introduced a few years ago (Gilb, 1973). This approach has proved useful in helping to detect certain types of "finding word" errors by comparing items within a set of data itself. Checkwords can be used for purposes other than detecting errors, for example, when attempting to find a record in a file. Like the check digit, the checkword is redundant information. Examples of checkwords are:

Example	*Explanation*
71132 BROWN	Employee number, with employee's last name
655T58 Legibi	Book call number, plus first six characters of title
655T58 LOP	Book call number, plus first letter of each word in title
70563 Mendham	Zip code, plus name of town
688-7829 RWB	Telephone number, plus customer's initials
6321 Shirt	Stock number, plus first five characters describing the item

Checkwords are extra descriptions provided by the designer as redundant information along with a basic code. This, obviously, can assist the computer in detecting more errors. Checkwords should never be derived from the original or basic code. That is, it is not a good idea to have the following combination:

Basic Code	*Checkword*
745-3216	73

The checkword "73" was derived from the basic code, $\underline{7}45\text{-}\underline{3}216 = 73$. The basic code and the checkword should come from totally different sources. Ideally, checkwords should be fairly short, possibly from one to four characters. However, some reasonable checkwords could be much longer. The content of the checkword is generally not too critical, as it is being used to confirm the accuracy of data, rather than as the primary source of information. In designing checkwords, it is usually a good idea to first make a list of the most plausible errors. It even may be a good idea to conduct a little study to see what kinds of errors people are making. Then determine which checkword combinations would help reduce these errors.

Limit Checks

Another computer-based detection capability includes "limit checking" (reasonableness checks). Many computer systems can make use of some form of limit checking. For example, a limit test may be as follows:

> If the number of items ordered is greater than *50* or less than *1,* generate an error message.

A designer should try to develop a means to check for reasonableness that makes full use of the computer's ability to "learn." It is well within the state of the art to develop limit checks that change as the computer gains experience. That is, the computer "learns" what the upper and lower limits should be as numerous orders are processed. You should provide for automatic changes based on the frequency of "out-of-limits" processing that occurs. For example, you may want to develop a limits check that lets 90% of all transactions pass automatically, another 8% to have a warning printed-out, and the final 2% held and reviewed before further processing can take place. Limit checking frequently is used to detect "large" errors, particularly where an extreme data item is entered into the computer, e.g., a $10000 is entered rather than a $100.00.

Pattern Checks

Computers also can detect errors by using "pattern checking" routines. If we assume that certain data fields will always have certain patterns of characters, we can have the computer check for that pattern. Examples include checking for all letters or all numbers, for even numbers or odd numbers, for multiples of 12 (dozens), for multiples of 5 or 10, etc.

Many computer error detection routines capitalize on the knowledge that some alpha characters and some numeric characters are frequently substituted across vocabulary lines; e.g., letters for numbers and numbers for letters. For example, consider the results from the two studies summarized below (Bailey, 1975; Owsowitz and Sweetland, 1965):

DIRECTION OF SUBSTITUTION	*STUDY 1*	*STUDY 2*
Letter for letter	25.8%	36.3%
Letter for number	49.0%	43.7%
Number for number	5.9%	3.2%
Number for letter	19.3%	16.8%
	100.0%	100.0%

Information from these studies indicates that from 60% to 70% of the character errors are substitutions across vocabulary lines; i.e., letter for number or number for letter. This means that if a error detection program was developed for detecting letters in number fields or numbers in letter fields, about 60% to 70% of this type error could be identified. For most data systems, however, this reduction is not generally achieved because many data fields do not require "all letters" or "all numbers". For this reason, it is generally difficult to use the "letter for number" or "number for letter" edits to identify single character substitutions.

Where it is possible to use this relatively simple and straightforward error detection scheme you should do it. But keep in mind that this or any other error-detection technique should *not* be used if it serves little or no useful purpose. Study the data fields and types of errors expected ahead of time so that computer resources are not spent on error detection techniques that buy you nothing. Remember that all error detection techniques have costs associated with them. This means you should always be in a position to estimate the relative advantages versus the costs of each error detection technique you include in a new system.

Restricting or eliminating the use of certain "look alike" characters helps to prevent errors. It also has a significant and positive impact on computer error detection capabilities. By using more selective character vocabularies, computer editing of input data no longer needs to rely so heavily on looking only for letters in number character positions, or numbers in letter character positions for detecting character substitution errors. With a restricted vocabulary the computer can also look for other illegal characters.

As a general rule, the fewer the total number of different characters used, the more efficient will be the character-level computer editing capabilities. Obviously, using the entire alpha-numeric character set (all 36 possible letters and numbers) substantially restricts the computer's capability to detect illegal characters. If only 20 different characters were used, and the computer was programmed to detect as errors any others that may be entered, the computer-detection capabilities are improved. Thus, when making up codes for a new system, if you don't use characters that "look alike" when reading or "sound alike" when listening it not only helps to *prevent* errors, but also makes it easier to *detect* them.

SOFTWARE ERRORS

We will briefly discuss what may be called "human errors once removed." A program usually will perform exactly as written, and if properly debugged and tested, there should be no program-based errors. However, large programs are sufficiently complex to harbor latent errors which may not show up for weeks, months, or even years after the program has been in operation. In one case, a large payroll program worked well for several years but failed when required to process name changes for two newly married female employees whose names were on adjacent records in the master file. The program was not able to handle this somewhat unusual situation. It is possible that all aspects of certain programs never may be fully used.

It therefore makes sense to put various error control features into the program. Among the types of error controls which can be used to test computer processing are the limit and reasonableness test, the crossfooting or crosstesting check, and control figures.

As with input data, a control over processing can be exercised by program steps testing the results of processing by comparing them with predetermined limits, or by comparison with flexible limits testing the reasonableness of the results. In a payroll application, the net pay can be checked against an upper limit. The upper limit is an amount where any paycheck exceeding the limit is probably in error. In a billing operation for certain types of products, such as coal or steel, the weight of the shipment may be divided into the billing rate in order to develop a price per pound. If the price per pound exceeds the average by more than a determined percentage, a message should be written for subsequent follow-up to determine if the billing is in *error*.

It is frequently possible to check computer data processing in a manner similar to the manual method of crossfooting. Individual items are totaled independently and then another total developed from the totals. For example, in a payroll application the totals are developed for gross pay, for each of the deduction items, and for net pay. The total for net pay is then obtained independently by taking the total for gross pay and deducting the totals for each of the deduction items. If this technique does not yield identical figures, there could have been an error in the program processing.

Finally, control figures can be used for testing the data processing within the computer. For example, the number of items to be invoiced in a billing run may be used as a control total and compared with the number of items billed on invoices.

CONCLUSION

As we have just reviewed, there are a number of techniques for detecting errors. System cost and complexity are increased by including measures which improve the accuracy of a system. Designers should evaluate the cost of improved accuracy (i.e., fewer errors) achieved by a particular error detected method against the benefits of the method. You should make some tradeoffs. We generally can't perform all the computer detections we would like to. So whatever error detection techniques you decide to use should have high payoff as far as system accuracy is concerned. The systems complexity and cost will increase as more computer-based error detection techniques are added.

EXERCISES

Proofread the following paragraphs as quickly as possible. After one proofreading of the paragraphs, check the list of errors at the bottom of this page to see which errors you failed to detect. Is there a pattern? What is the field-level error rate? What would have been the reported field-level error rate if you counted only the errors you detected?

Suggest ways of having a computer detect many of the errors, particularly those that are difficult to catch by proofreading. Report your findings and discuss your analysis in a one-page report.

How bad is the medication error problem in hospitols? Medication error studys suggest that it is about 12%. If we project a 12% medication error rate for a hospitol with an average lode of 200 patience, and assume that each patient will recieve 10 doses a day, we would expct about 2400 errors to be made daily. But some people are skeptical of these figures, and think the real error rate is somewher around 1%. Such an error rate would still mean that we would have an avvrage of about 2 errors daily. If these errors are serious enough, it could mean two deths daily.

In a recent study completed in 1880, over 200,000 doses were adminstered, but incident reports were submitted for only only 36 errors. Observations of ersors showed that the error rate was at least 12%, which means that about 240 reports should have be been prepared.

Obviously, in a hospitol where each pateint is given about 10 dailey doses ordered by physicians, filled by physicians, and administered to patients by nurses, errrors will occur.*

* The above three paragraphs contain 180 words and 23 field-level errors. The errors and correct responses, shown in the order they appear in the paragraphs, are: hospitols-hospital, studys-studies, hospitol-hospital, lode-load, patience-patients, receive-recieve, expct-expect, 2400-240, somewher-somewhere, avvrage-average, 2-24, deths-deaths, 1880-1980, adminstered-administered, only only-only, ersors-errors, 2,400-24,000, have-had, be been-been, pateint-patient, dailey-daily, physicians-pharmacists, and errrors-errors.

13

ERROR CORRECTION

INTRODUCTION

In 1874, an attorney, involved in a troublesome patent case, described a great moment in the history of invention before the Supreme Court of the United States:

> Lead pencils have very long been used to make marks. India rubber has very long been used to rub them out. But never until lately was India rubber used for this purpose except in a form *disconnected* from the pencil. But on a summer's morning of 1867, one J. B. Blair, a poor artist of Philadelphia, seeing that it would be more convenient to use the rubber *on* his pencil rather than off, put a piece of India rubber on his pencil. Behold ... hundreds of thousands and millions of rubber-tipped pencils at once appeared (New York Times Magazine, April 27, 1975).

The attorney was obviously defending the invention of the lead pencil-and-eraser combination. Unfortunately for Mr. Blair, a H. L. Lipman had been nine years previously awarded a patent for the pencil with eraser. Even so, in the latter part of the 1800's a convenient eraser had come into being to assist those who needed to more quickly correct errors.

For many years thereafter, correcting most data errors was a simple matter of erasing. Even today, the erasure is frequently used to correct errors (including those made by pencils *and* pens). Most modern typewriters are now equipped with a special means for correcting errors. Computer-based word processing systems (text editors) usually offer several ways to correct errors. We have progressed to the point where making many types of corrections is getting easier and easier.

As mentioned in the previous chapter, a good system designer will include a series of appropriate and effective self-detection, proofreading and computer-detection capabilities. In addition, a good designer also will provide a workable means for efficiently correcting the errors that are detected. It is a big mistake to leave these important design considerations to a group of new and inexperienced users. System designers have the responsibility to "think through" and prepare sound ways for each and every error to be rapidly corrected.

INPUT MODES

There are two modes to be considered when discussing error correction: the batch mode and the interactive mode. The batch mode includes all situations where a user enters information to a computer, and at least part of the computer-detection process takes place a considerable time (possibly several hours) after it is entered. In the interactive mode, all computer detection of errors takes place while information is being entered. This usually means that errors in the interactive mode are corrected right after they are detected.

MAKING BATCH MODE CORRECTIONS

Many batch oriented error correction facilities can best be described as "dump and run". Errors are detected by the computer and literally dumped on users. In many of these situations, designers have given little thought to what needs to be done with the errors in order to get corrected data back into the system. Many users have stood by helplessly as "errors" spewed forth in large numbers, with little or no good way provided by the system designer to make corrections and quickly get the corrected data back into the computer. Believe it or not, some systems actually have failed because large numbers of computer-detected errors could not be corrected fast enough.

Studies have shown that many designers have a general lack of understanding of issues related to error correction. For example:

1. Designers tend to underestimate the *number* of errors that will need correcting (in one system they estimated a 5% error rate that turned out to be 20%);
2. Designers tend to underestimate the *time* it will take to correct each error (in one system they planned on having each clerk correct 100 each day, but even the best clerks were never able to correct more than 25);
3. Designers do not realize that certain types of errors account for the majority of time spent in correcting errors (in one system, about 71% of a clerk's time was spent in correcting just 3% of all errors);
4. Designers do not realize that the correction (and re-entry) process itself creates errors (in one system about 25% of corrected entries were wrong);
5. Designers do not realize that some errors are more critical to system success and require special handling;
6. Designers do not realize that it is usually a small number of errors that tend to occur over and over again, while some potential errors never occur (in one system, designers developed over 1200 different error messages, but only about 200 were ever used).

THE COSTS OF CORRECTION

The ways and means that people use to correct errors in computer systems is very important. Faulty design decisions concerning error correction can as much as double or triple the number of people needed to operate a new system. A few years ago many experts were predicting that with the use of computers many people would lose their jobs and we would have terrible unemployment problems. Many workers actually resented new computer systems being introduced into their companies because they thought they would be replaced. For the most part, both experts and workers were wrong. Computers have not contributed nearly as much to unemployment as was predicted.

One of the major reasons that more people have not lost their jobs is that computers continue to "demand" correct data. The more errors that are detected, the more errors that have to be corrected. In many organizations (if not most), workers have been shifted from making tedious calculations, manual updating, and endless filing, to *correcting errors.* In fact, much of the most costly human performance in many new computer systems comes from people correcting errors.

Error Messages

In order for users to efficiently correct errors, system designers should provide:

(a) Error messages

(b) Error correction strategies, and

(c) Error re-entry procedures.

Once the computer has detected an error, a clear, concise and complete error message should be made available to the error corrector. As a minimum, an error message should include (a) the *error* (underlined or highlighted in some way), (b) an *explanation* as to why the error was detected, and (c) a *suggestion* of how to correct the error. These three items of information are highly desirable a for good error messages in most systems.

In batch systems it is usually a good idea to include a copy of the entire input record. In addition, it is also helpful to include (a) the date and time the error was detected, (b) the name, initials, or number of the person who made the error, and (c) references where a person can look to find more information concerning how to correct that type of an error. Incidentally, the date and time that an error was detected is included on an error message to help control and expedite the correction of errors (the sooner corrected, the better).

Also in batch systems, if at all possible the error message should be sent directly to the error-maker. The error-maker should be required to make all needed corrections in a timely fashion. Besides having the person who "slept in the bed have to make the bed", this policy also provides error-makers with timely feedback about the number and kinds of errors they are making. In addition, the error-maker probably has the best chance of correcting the error in the shortest period of time. The less time spent correcting each error, the more errors each person can correct.

As noted earlier, by making their own corrections, users will be constantly aware of the numbers and types of errors they are making. If expected to correct their own errors, users will have less of a tendency to rush through their work with little thought of accuracy. If they know that someone else will "clean up after them", users are more likely to continue with unreliable, system degrading performance. As a general rule, each error should be corrected by the person who made it.

Remember that another advantage of having error-makers correct their own errors is that they frequently have the most complete information as to why the error occurred, and what can be done to correct it. As errors are corrected further and further from their origin, they become more difficult and time-consuming to fix. For example, a billing error that is self-detected and immediately corrected costs far less then correcting the same error found by a customer 30 to 60 days after it was made. Also as a general rule, as errors continue undetected from their source, they tend to be related to other errors occurring. This means they become more and more of a hindrance to successful system operation.

ERROR CORRECTION STRATEGIES AND PROCEDURES

For each error message, the error-corrector should be provided with detailed instructions including a strategy, for quickly and efficiently correcting the error.

Not all errors require the same effort to be corrected. When the need arises to correct an error, it may be a very simple matter of backspacing and typing over, or a very long, tedious and involved process. With difficult corrections, this typically means first comparing the conflicting information, and then tracing back to the point where a record first came to be in error. This may involve checking on information sources that were created long before the system was designed; sometimes going back to entries dating decades earlier (e.g. security transactions, estate settlements leases, running accounts of many kinds, etc.). In some large-scale clerical operations this kind of activity is referred to as "researching the error". It may occupy the full-time attention of one or scores of employees. You need to keep in mind that some errors can be corrected very quickly, while others can require considerable error correction research.

Among other things, the time it takes to correct an error depends on both the complexity of the error and skill level of the error corrector. Error-correction clerks are generally best qualified to correct *clerical processing* errors. Their ability to deal with problems of ignorance or misunderstanding on the part of error-makers is usually much more limited.

A system should be designed so that errors can be corrected in the shortest possible time frame. If a clerical force spends an average of 20 minutes to correct each error rather than an average of ten minutes, about twice the number of clerks will be needed for correcting errors. By adding more people to make a new system "work", it obviously costs substantially more to operate, which reduces whatever cost benefits promised for the new system. The one place where we can "unexpectingly" add the most people the fastest is in helping to correct errors.

All errors detected by a computer should be kept in one or more "error files" until corrected. In batch systems particularly, the file should contain as a minimum, the error, the reason it was detected as an error (i.e., the computer process for detecting it), and the date it was detected. As mentioned earlier, errors that are detected, always should be corrected as soon as possible. In a batch mode it is obviously not practical to stop the computer processing each time an error is detected and wait for it to be corrected before continuing. But it is equally ridiculous to detect an error and then shunt it into an error file where it remains uncorrected for days or even weeks. Ideally, all computer detected errors would be placed in a temporary error file, where they would be corrected within 24 hours.

One way to save time in correcting errors is to place them directly in the hands of the people most capable of resolving them. A second means of saving time is to sort the errors by nature of the problem. Sorting by problem usually must be conducted by individuals thoroughly versed in the complexities of the total system. These people should be aware of the usual procedures for resolving discrepancies. The errors then can be given to people to correct according to the ability level of the error corrector.

ERROR RE-ENTRY PROCEDURES

For all errors there should be a single, straight-forward means of entering the corrected information back into the computer. The system designer should make it a policy to keep as much of the original (correct) data in the computer as possible. We should not make the mistake of requiring error-correctors to re-enter a large number of data fields just to get *one* corrected field back into the computer. The more data that has to be re-entered, the greater the probability a *new error* will be made. Keep the amount of data to be re-entered, and the re-entry procedures as simple, straight-forward and brief as possible. It is worthwhile to repeat that if a system is designed so it requires extensive re-entering of data with each corrected error, a considerable number of new errors will be made.

Many of the same considerations that are valid in the batch mode are also valid in the interactive mode. Because errors are usually corrected right after they are detected, it is important to provide convenient ways to make the corrections.

Possibly the most important rule is to make the error correction procedure as simple as possible. Consider the following example:

Computer: How many codes do you want to enter?
User: 13

Computer:	Enter code 1
User:	72634
Computer:	Enter code 2
User:	A38416
Computer:	Enter code 3
User:	52B
Computer:	Enter code 4
User:	P38D84
Computer:	Enter code 5
User:	X7R
Computer:	Enter code 6
User:	S
Computer:	Enter code 7
User:	36780W
Computer:	Enter code 8
User:	CB-26R
Computer:	Enter code 9
User:	73166
Computer:	Enter code 10
User:	32d
Computer:	*****ERROR IN CODE 10 -- LOWER CASE LETTERS INVALID
Computer:	How many codes do you want to enter?

In this situation the designer required the user to enter all of the information twice. Not only is this frustrating and takes a lot of extra time, but it also provides a good opportunity for errors to be made in another field. The computer should save everything that already has been entered. The user should be required to correct the error by making the fewest keystrokes possible. The error-correction procedures should be easy to learn, remember and implement.

The ideal situation is to have error correctors re-enter the corrected data and possibly *one* other identifying field (if necessary). If the error message is a printout, we can design the printout to be used as a "turn-around" form. That is, it serves double duty, by informing a clerk about the error and then by being used as an input form. This reduces the amount of handwriting a person must do to prepare the corrected data for input, which saves time and reduces the number of new errors made.

To summarize, the three major ingredients of efficient error-correction include knowing why it was detected, understanding what needs to be done to fix it fast, and being able to quickly and accurately re-enter the correction.

COMPUTER CORRECTION

Probably the biggest advantage of having the computer correct errors is that it saves the time and expensive of having people carry out the correction process. Two brief examples of having the computer make corrections will be presented. These should help illustrate the methods that are available.

A computer can be programmed to automatically *correct* many handprinting, typing and spelling errors that it detects. These types of errors will be referred to as "clerical" errors. Work on automatic spelling correction was first reported several years ago (Glantz, 1957), and has been improving ever since. Many computer systems have been devised to make corrections. Moreover, it is not particularly costly to use the computer to correct many types of clerical errors.

Efficiency with correcting clerical errors best can be obtained by limiting the types of errors to be corrected. In other words, don't start out trying to computer-correct all detected errors. Begin by correcting those that are easiest.

Clearly, the types of clerical errors which can occur are many in number, and designing a system to correct all of them would be a difficult if not impossible task. Fortunately, about 80% of all clerical errors seem to fall into one of four classes:

(a) one character is wrong,

(b) one character is omitted,

(c) an extra character is added,

(d) two adjacent characters are transposed.

By limiting the class of errors to be corrected to the above classes, we have effectively reduced the number of items to be dealt with.

Most of the work on correcting these types of clerical errors has focused on the general problem of recognizing wrong words in a string of plain English text by using a dictionary of "correct" spellings. Since only single letter errors and transpositions are considered, any dictionary words whose lengths differ by more than one letter from the entry word may be detected (and corrected) immediately.

In many cases, once an error is corrected the potential correction is shown to a user for verification. This can be done either at a terminal in an interactive mode, or on a printout in a batch model. The user can then accept or not accept the correction.

CORRECTING HANDPRINTING ERRORS

If most of the data coming to your new system is from handprinted input, you should consider having the computer correct many of the errors it detects. By using the tables in Appendix E, it is possible to develop a computer program to correct many of the codes that have been detected by the computer as incorrect. This example is provided to illustrate one way of using the computer to correct errors.

Remember, when using the tables in Appendix E, the input must be primarily from *handprinted* characters. This approach only will work effectively with keying errors or speaking errors if a different set of tables are used. Also, this technique works best when used with codes that are incorrect because of a character substitution error.

Mixed Vocabulary Errors

The first step for automatically correcting handprinting errors is to determine the character vocabulary originally used to construct the code. This could be all alpha (only letters), all numeric (only numbers) or mixed alpha-numeric (both letters and numbers).

The next step is to determine which characters are most likely to have been substituted for the right character. Frequently, computer-detection of an incorrect code also involves the identification of the incorrect character. However, if the specific character is *not* identified as part of the detection process, than another approach must be used to identify the wrong character. This can be done by using the information contained in Appendix E:

(a) Table E-1 should be used for a mixed alphanumeric character vocabulary,

(b) Table E-2 for an all alpha character vocabulary, and

(c) Table E-3 when an all numeric vocabulary has been used.

For example, assume that the computer determined that the code U34WL was wrong. To make the corrections assuming a mixed alpha-numeric character vocabulary was used to originally construct the set of codes (i.e., all 36 characters), we need to determine which character most likely has been substituted? The breakout would appear as follows (from Table E-1):

Code Characters	*Total Substitution Rate*
U	.193
3	.028
4	.040
W	.109
L	.026

From this analysis, it appears that the character that is the most likely substitute is the "U", second most likely is the "W", and the least likely is the "L".

The next step is to determine for each character the most likely substitution. For the character "U", the following substitution options are most likely (also from Table E-1):

Character	*Substitution Rate*
V	.134
4	.028
0	.006
Q	.006

From our analysis so far, it appears that "U" was most likely the wrong character, and it was most likely substituted for "V". However, "U" could have been substituted for "4", "0" or "Q", but the probability of this is much lower. Thus, the correct code is most likely *V*34WL rather than *U*34WL.

If several alternative corrects are to be presented to users, the analysis could continue until all highly probable substitutions are taken into account. These other possibilities are shown below:

Code Characters	*Possible Substitutions & Substitution Rates*			
U	V (.134)	4 (.028)	0 (.006)	Q (.006)
3	B (.006)	J (.006)	S (.005)	
4	Y (.027)	A (.005)		
W	N (.095)			
L	None			

Note that the character "W" is substituted for the character "N" quite frequently. Thus, the second most likely correct code is U34*N*L. The third and fourth alternatives are *4*34WL and U3*Y*WL. The results of the entire analysis, which could be presented to users, are summarized below:

Incorrect Code	U 3 4 W L
Correct Code-1st Alternative	V̲ 3 4 W L
Correct Code-2nd Alternative	U 3 4 N̲ L
Correct Code-3rd Alternative	4̲ 3 4 W L
Correct Code-4th Alternative	U 3 Y̲ W L

Even though only four alternatives were developed, there are actually ten possible "corrections" for the incorrect code used in the above example. The latter six alternatives are all about equally probable, and each of the latter six are much less probable than the first four alternatives.

If you desire to have the computer make the correction without user intervention, the computer would compute "corrections" and recycle the code back through the error-detection routines until an acceptable code is found.

Single Vocabulary Codes

Another example, that is slightly different, consists of a situation where an all numeric or all alpha code vocabulary is used. When the incorrect code is all numbers, e.g., 21345, the same process is used that is described above. But when an all numeric code is wrong because it contains an alpha character, e.g., 2*H*345, the correction problem is simplified. In this case, the incorrect character is usually quite obvious. Whenever an alpha character is substituted for a numeric character, or a numeric character is substituted for an alpha character, the first (and sometimes only) step is to determine the most likely direction of substitution. For the code 2H345, the correct code is most likely 21345. In fact, there are no other reasonable alternatives.

When using an all alpha or all numeric vocabulary, if most codes contain only one incorrect character and if 87% of all character substitution are in a consistent and predictable direction (Bailey, 1975), it seems that with proper programming a considerable number of these errors can be detected *and corrected* by the computer. Based on a statistical analysis of the data collected in one study, it is suggested that from 50% to 70% of the codes that are wrong because of handprinting errors, and that are *detected* by the computer, also can be accurately *corrected* by the computer. This somewhat optimistic prediction is due primarily to the fact that very few character combinations account for most substitution errors.

EXERCISES

Consider the errors shown in the Exercise at the end of Chapter 12. Develop a way for a computer to detect and *correct* many of these errors *as they are being made.* Assume that a person is entering the paragraphs using a computer-based text editor. Propose a means for doing this that seems technically feasible, and that will not irritate users. Describe your ideas in a one-page report. Be creative!

APPENDICES

APPENDIX A

HANDPRINTING ERRORS

A listing of all 36 alpha and numeric handprinted characters ranked by lowest error rate to highest.

CHAR-ACTER	*ERROR RATE*	*CHAR-ACTER*	*ERROR RATE*
W	.44%	B	1.73%
M	.53	T	1.86
3	.53	U	2.00
7	.67	P	2.22
A	.84	S	2.22
9	.89	L	2.40
E	.89	Q	2.86
C	.98	Y	2.93
8	1.02	4	3.91
O	1.06	1	4.13
K	1.16	N	4.53
R	1.20	0	4.98
F	1.24	5	5.02
X	1.38	J	5.06
H	1.42	V	5.15
6	1.42	G	5.64
D	1.60	Z	13.33
2	1.69	I	24.62

APPENDIX B

READABILITY MEASURES

Although long sentences using a difficult vocabulary with multi-syllable words may be preferred by certain readers, most users in systems prefer instructions made up of *short sentences* with familiar words of *one or two syllables.* Readability measures have been developed to assist designers in preparing instructions that make better use of word and sentence characteristics.

Readability measures provide a means for estimating the difficulty a reader may have with a set of instruction. In using a readability measure to evaluate instructions, a decision has to be made as to what constitutes an acceptable level of readability for the material being prepared. This decision should be based on information about the reading abilities of the intended audience. As a general rule, it is better to write to a readability level that is *below* the reading skills levels of the intended audience. The reading ability of a group of potential users can be measured with a standardized reading test (cf. Nelson-Denny).

If it is not practical to use a reading test, some idea of reading skill levels can be obtained by looking at the user's educational levels. In general, people with more education have better reading skills than people with less education.

The readability measures that follow use only two features of text to predict readability: a measure of word difficulty, and a measure of sentence difficulty. The first is the Kincaid (Kincaid, Fishburne, Rogers and Chissom, 1975), which uses average length of a word in syllables to measure word difficulty. The second will be the Automated Readability Index (Smith and Kincaid, 1970), which uses the average number of letters per word to measure word difficulty. Both use the average words per sentence as a measure of sentence difficulty. When a formula is being computed by hand, it is easier to use the Kincaid where syllables are counted rather than letters. When a formula is being computed by a computer, letters are easier to count than syllables, which is what the Automated Readability Index does.

To calculate the readability of a document using the Kincaid formula,* a designer should:

* The process is almost the same for using the Automated Readability Index. The only difference is that with the Automated Readability Index the average number of letters per word is used rather than the average number of syllables per word. Using the average number of letters per word facilitates the use of this formula with computers. The Automated Readability Index formula is shown below for designers who want to computerize the readability calculation process:

1. Select five or more 100 to 150-word samples,
2. Determine the average number of syllables per word in each sample,
3. Determine the average number of words per sentence in each sample, and
4. Apply the following formula to make the calculation:

$$\text{Reading Grade Level} = 11.8S + 0.39W - 15.59$$

where S = average syllables per word
W = average words per sentence.

Abbreviations, acronyms and numbers often can be a problem when syllable counts are made because these words usually do not conform to the usual rules for determining syllables. When counting syllables, drop all words whose pronunciation is in doubt. Also, define a sentence as a set of words terminated by one of the punctuation marks commonly used to end sentences (e.g., period, question mark or exclamation point). As an example of how to use the Kincaid readability formula, consider the following paragraph. It is a 127-word section from a much larger document.

> Readability formulas predict reading difficulty, but they do not reveal why a document is difficult. Many important attributes of style that contribute to difficulty cannot be quantified. These unmeasurable attributes are often highly correlated with stylistic features that can be measured. For example, a poorly organized document might contain a lot of long sentences because the writer is forced to reference topics he or she has or will discuss. In this case, the document would be classified as difficult by most readability formulas since average sentence length is often used in these formulas. But shortening sentences to produce a more "readable" document would not get at the source of difficulty. A readability formula should be used as a *predictor* of difficulty and not as a *diagnostic tool.*

This selection contains 230 syllables. The number of syllables (230) is divided by the total number of words (127), which produces an average

---

$$\text{Reading Grade Level} = 4.71\ L + 0.5\ W - 21.43$$

where L = average letters per word
W = average words per sentence

syllables per word of 1.81 ($S = 1.81$). The 127 words are contained in 7 sentences. The number of words (127) is divided by the number of sentences (7), which produces an average words per sentence of 18.14 ($W = 18.14$). These results are then used to enter the formula:

$$
\begin{aligned}
\text{Reading Grade Level} &= 11.8\ S + 0.39\ W - 15.59 \\
&= 1.8 \times 1.81 + 0.39 \times 18.14 - 15.59 \\
&= 21.36 + 7.07 - 15.59 \\
&= 12.84
\end{aligned}
$$

The reading grade level for this one section is 12.84. A designer should take similar readability counts for at least four other sections (the more the better) and then average them. For example, assume that the following readability counts were taken for a document:

Section	*Readability Count*
1	12.84
2	11.03
3	10.65
4	13.48
5	14.72
6	11.64
7	9.87
8	13.52

The average of these counts, which is 12.22, will be the readability score or reading grade level for this document.

APPENDIX C

INPUT DEVICE GUIDELINES

The following list of human performance requirements can be used when making an initial evaluation of a video display terminal—a CRT/keyboard combination. Terminals that meet other important system requirements can then be compared from a human performance point of view. Each terminal can be evaluated on each of the human performance requirements. When a requirement is met the weighted value can be added to the total of other "met" requirements. The weights give an estimate of the relative importance of each requirement. The weights range from 1 to 5, with 5 indicating the most important requirements. At the end of an evaluation add the totals for the terminals and consider the one with the highest total as best meeting the human performance requirements.

CRT DISPLAY	*WEIGHT*
Location and Design - The CRT screen should be designed and located so that it may be easily read by users in their normal operating position.	5
Viewing Distance - A 16 to 24-inch viewing distance (eye to screen) should be provided. Design should permit the user to view the scope from as close as 12 inches.	4
Orientation - The CRT screen should be perpendicular to the user's *normal* line-of-sight. The angle of the CRT from the verticle should approximate '10 sup o'.	4
Reflectance - Displays should be constructed, arranged, and mounted to minimize or eliminate reflectance of the ambient illumination from the glass or plastic cover. Reflected glare should be minimized by proper placement of the scope relative to light sources, use of a hood or shield, or optical coatings.	4
Flicker - The CRT screen should have no discernible flicker and be free of any movement.	4
Screen Contrast - Contrast on the face of the CRT (with light characters against a dark background) should be between 3:1 to 15:1, recommended between 6:1 to 10:1 (ratio of brightest illumination to lowest illumination) in room ambient light levels of 70 to 100 ft-c. When dark characters are displayed against a light background, the luminance of the character background should be at least three times the luminance of the character.	3

Adjacent Surfaces - Surfaces adjacent to the scope should have a dull matte finish. The brightness range of surfaces immediately adjacent to the CRT should be between 10 percent and 100 percent of screen brightness. No light source (warning lights, mode indicators, etc.) under control of the vendor in the immediate surround should be brighter than the CRT characters. 1

Operator Controls - A limited range CRT brightness control and a contrast control should be provided for the user. All other controls—linearity, height, width, and focus—should be provided, but accessible only to maintenance personnel. 2

Cursor - The equipment should have a character underline, motionless cursor. The cursor should have the following characteristics: left, right, up, down. 5

Character Height - Characters should be at least 0.17 inches in height. 4

Character Width - Character width should be between 70 percent and 85 percent of character height. 3

Stroke Width - Stroke width-to-height ratio should be 1:8 to 1:10. 3

Character Spacing - Spacing between adjacent characters should be at least 18 percent of the character height. 3

Interline Spacing - Distance between lines should be at least 50 percent of character height. A distance of 66 percent is preferred. 3

Character Resolution - The number of scan (raster) lines per character should be a minimum of 10. 4

Character Illumination - Characters should be evenly illuminated. 1

Format - Editing and error correction information should be presented to the operator in a directly usable form. Requirements for transposing, computing, interpolating or mental translation into other units should be avoided. 3

Content - The information displayed on the CRT screen should be limited to that which is necessary to perform specific actions or make required decisions. 3

KEYBOARD

Key Size - The key tops should be square or slightly rounded and have a diameter of 0.5 inches. 4

Key Surface - The key tops should be concave and treated to minimize glare.

2

Key Displacement - Displacement should be a minimum of 0.125 inches to a maximum of 0.187 inches.

3

Key Resistance - Resistance should be in the range of 2 to 4 ounces.

3

Key Labeling - Key symbols should be etched to resist wear, and colored with high contrast lettering.

1

Key Separation - The adjacent edges of keys should be 0.25 inches apart.

4

Key Mounting - Keys should be securely mounted and firmly fixed in place to minimize horizontal movement.

2

Keyboard Angle - The keyboard should be inclined at an angle of ‘11 sup o‘ to ‘15 sup o‘ from the horizontal.

1

Keyboard Height - The "home" row of the keyboard should be about 30 inches from the floor.

3

Highlighting - Functional highlighting of the various key groups should be accomplished through the use of color coding techniques.

2

Special Keys - Special keys should be an integral part of the keyboard. The distance between them and the alphanumeric section of the keyboard should not exceed 1 inch.

3

WORKSPACE

Work Surface Width - A workspace at least 30 inches wide and 20 inches deep should be provided to accommodate records being transcribed.

1

Work Surface Height - The work surface top should be 30 inches above the floor.

3

Copy Holder - A copy holder should be available that can be freely adjusted by the user, and has a surface area larger than the documents being transcribed.	3
Palm Rest - A palm rest should be provided for the key operator.	3
Leg Room - Unobstructed leg room should be provided.	2

ENVIRONMENTAL FACTORS	
Noise Generation - Noise generated by the video display terminal should be less than the ambient noise level of the environment before the device is installed (about 50 to 60 dB). Other factors, such as continued highpitched whir or hiss also should be eliminated from the equipment.	3
Heat Generation - Heat generated by the equipment should be dissipated in a way that does not expose the user to excess heat either directly or through an uncomfortable increase in ambient room temperature.	3
Ambient Light Level - Provision shall be made for the satisfactory viewing of the CRT in a room having an ambient light level from 70 to 100 ft-c.	3

APPENDIX D

CODE DESIGN PRINCIPLES

Code Length

1. Make codes short.
2. Codes of eight or more characters should be broken into subunits of three and/or four characters (e.g., 543-2338).
3. Errors are positively correlated with increasing code length. Therefore, cost trade-offs between preventing errors and catching/correcting errors should be carefully considered before codes are artificially lengthened to provide for automatic checks (e.g., check digits).

Code Make-Up

1. Where the code-use task requires complex internal processes such as amplifying or synthesizing several codes into a meaningful code set, it is desirable and necessary to *build meaning* into the code.
2. Where the code-use task is simple, and codes can be short, such as keying unrelated codes from look-up, meaning is desirable but not necessary.
3. Since it is easier to convey meaning with alpha characters than with numerics, tasks requiring complex internal processing of the code should use alpha to the extent possible.
4. Mnemonics facilitate learning and should be designed into code sets where users are expected to learn any appreciable number of codes.
5. Where tasks involve only simple transfer of discrete codes, where a single code system is sued, and where the code system is sufficiently small to permit use of a single vocabulary, numeric codes should be used in preference to random alpha.
6. Codes that are to be extensively used should be composed of characters drawn from vocabularies already within the repertoire of users.
7. The code vocabulary for a given code set should be as small as possible consonant with unique coding and short codes, since the larger the vocabulary base from which a code is drawn the more difficult it is to remember the code.
8. When required size of vocabulary permits, characters which are easily confused perceptually with other characters should be omitted.

9. When use of characters with high probability of visual or auditory confusability is needed in code vocabularies, distinguishing features should be emphasized to make differences more readily evident.
10. Few users work exclusively with one code system. There is evidence that users integrate best that vocabulary type which dominates the code systems used by those users (i.e., alpha, numeric or mixed alpha-numeric). However, in tasks of simple transfer of discrete codes, where learning of codes is neither necessary to nor probable in the job context, and where the code system is sufficiently large to require both alpha and numeric characters, no great penalty can be expected by favoring numerics over alphas in the expansion of large code systems.
11. Where size of inventory being coded permits, appreciable error reduction can be attained by construction of single vocabulary code systems (i.e., either all alpha or all numeric). However, gains attained by suing one vocabulary will progressively disappear as larger inventories force code lengths to extend beyond eight characters.
12. The use of like symbols for the same function, e.g., the use of a dash or a virgule as a separator in a code phrase such as month-day-year, should be used consistently.

General (Applies in Design of All Codes)

1. For any given class of codes, rules for code design should be clearly stated and consistently applied.
2. Interdepartmental consistency in the construction of code phrases, e.g., the ordering of code units within a standard message unit such as month-day-year, is critical to the design of codes for efficient human performance.

For a more detail discussion of code design principles see Chapter 16 in Bailey, 1982.

APPENDIX E

COMPUTER CORRECTION TABLES

Table E-1. A listing of all 36 letters and numbers showing substituted character, stimulus character and substitution rates.

SUBSTITUTED CHARACTERS	*STIMULUS CHARACTER AND SUBSTITUTION RATE*	*TOTAL SUBSTITUTION RATE*
A	Q (.011) H(.005)	.028
B	8 (.005)	.021
C	G (.059) F (.012) 6 (.010) E (.007) 0 (.007)	.133
D	P (.033) B (.027)	.069
E	F (.010)	.020
F	I (.007) E (.005)	.030
G	6 (.016) C (.009)	.045
H	N (.009) K (.006)	.025
I	1 (.068)	.078
J	S (.010) I (.006)	.028
K	X (.015)	.033
L		.026
M	N (.006)	.020
N	M (.007) R (.007) H (.005) W (.005)	.041
O	Q (.240) 0 (.040) U (.006)	.312
P	D (.012)	.025
Q	O (.059) O (.009) G (.005)	.083
R		.016
S	5 (.126) J (.051) 8 (.009)	.204
T	J (.006)	.016
U	V (.134) 4 (.028) O (.006) Q (.006)	.193
V	U (.030) Y (.015) X (.010)	.064
W	N (.095)	.109
X	Y (.021)	.031
Y	4 (.059)	.075
Z	2 (.028)	.035
0	Q (.007)	.011
1	I (.650) L (.051) H (.011)	.728
2	Z (.347) 1 (.021) 7 (.006)	.385
3	B (.006) J (.006) S (.005)	.028
4	Y (.027) A (.005)	.040
5	J (.060) S (.027)	.098
6	G (.073) O (.011) O (.006)	.109
7	T (.037) 1 (.016) D (.010) Z (.009) 9 (.006)	.096
8		.012
9	P (.006) G (.005) 4 (.005)	.031

TABLE E-2. A listing of 26 letters showing substituted characters, stimulus characters and substitution rates when using only alpha characters.

SUBSTITUTED CHARACTERS	*STIMULUS CHARACTER AND SUBSTITUTION RATE*	*TOTAL SUBSTITUTION RATE*
A	Q (.015) H (.007)	.030
B		.019
C	G (.082) F (.017) E (.010) U (.009) L (.007)	.150
D	P (.046) B (.039)	.087
E	F (.014)	.021
F	I (.010) E (.007) P (.007)	.032
G	C (.012)	.027
H	N (.012) K (.009)	.029
I		.009
J	S (.014) I (.009)	.026
K	X (.021)	.041
L		.022
M	N (.009)	.024
N	M (.010) R (.010) H. (.007) W (.007)	.056
O	Q (.333) U (.009)	.369
P	D (.017)	.027
Q	O (.012) G (.007)	.029
R		.019
S	J (.070)	.080
T	J (.009)	.014
U	V (.186) O (.009) Q (.009)	.224
V	U (.041) Y (.021) X (.014)	.084
W	N (.132)	.147
X	Y (.029)	.038
Y		.019
Z		.003

Table E-3. A listing of 10 numbers showing substituted characters, stimulus characters and substitution rates when using only numbers.

SUBSTITUTED CHARACTERS	*STIMULUS CHARACTER AND SUBSTITUTION RATE*	*TOTAL SUBSTITUTION RATE*
0		.004
1		.022
2	1 (.076) 7 (.022)	.120
3		.009
4		.018
5		.022
6	0 (.040)	.044
7	1 (.058) 9 (.022)	.102
8		.009
9	4 (.018)	.022

REFERENCES

AT&T, War On Errors: A System Task Force Study in Indianapolis, Bell System Report, October, 1966.

AT&T, Assignment office quality control study, Bell System Report, 1967.

Alden, D. G., Daniels, R. W., and Kanarick, A. F., Keyboard design and operation: A review of major issues, *Human Factors,* 1972, *14*, 275-293.

Altman, J. W., Improvements needed in a central store of human performance data, *Human Factors,* 1964.

Altman, J. W., Human error analysis, Paper presented at Quality Control Conference, American Society for Quality Control, October, 1965.

Anderson, R. W., The physicians contribution to hospital medication errors, *American Journal of Hospital Pharmacy,* 1971.

Bailey, R. W., A classification scheme for human error in computer-based business information systems, Bell Laboratories Report, May, 1972.

Bailey, R. W., The reliability of human performance, Bell Laboratories Report, February, 1973.

Bailey, R. W., Handprinted characters and human error, Bell Laboratories Report, December, 1974.

Bailey, R. W., Human Performance Engineering: A Guide for System Designers, Englewood Cliffs, N.J.: Prentice-Hall, 1982.

Bailey, R. W. and Desaulniers, D. H., Directory error analysis, Bell Laboratories Report, March, 1978.

Bailey, R. W., Blank, R. G. and Walker, J. T., A Study of human error in the operation of the TIRKS F/ECP-1 data system, Bell Laboratories Report, July, 1975.

Barker, K. M., The effects of an experimental medication system on medication errors and costs, Part one: Introduction and errors study, *American Journal of Hospital Pharmacy,* 1968.

Barone, F. L., Eisner, W. J., Galli, E. J., FTD semi-automatic file conversion system, *Technical Report,* AF 30(602)-2860, March, 1965.

Carlson, G., Predicting clerical error, *Datamation,* February, 1963.

Chapanis, A., Words, words, words, Presidential Address at Eighth Annual Meeting of the Human Factors Society, October 6, 1964.

Chesapeake and Potomac Telephone Co., How Accurate Is the Service Order Accuracy Report? Bell System Report, January, 1967.

Chession, F. W., Computers and Cryptology, *Datamation,* January, 1973.

Conrad, R., Errors of immediate memory, *British Journal of Psychology,* November, 1959, 50, 349-359.

Conrad, R., Acoustic confusions and immediate memory, *British Journal of Psychology,* 1964, 55, 75-84.

Conrad, R. and Hull, A. J., Copying alpha and numeric codes by hand, *Journal of Applied Psychology,* 1967, 51, 444-448.

Cornell, C. E., Minimizing human errors, *Space/Aeronautics,* March, 1968.

Deininger, R. L., Human factors engineering studies of the design and use of push-button telephone sets, *Bell Systems Technical Journal,* 1960, 39, 995-1012.

Devoe, D. B., Alternatives to handprinting in the manual entry of data, *IEEE Transactions on Human Factors In Electronics,* March, 1967.

Dreyfuss, H., *Anthropometric Data,* New York: Whitney Publications, 1959.

Elliot, A., A study of errors in skilled performance, *Unpublished Thesis,* University of London, 1959.

Ford, R. N., *Motivation Through Work Itself,* American Management Association, Inc., 1969.

Ford, R. N., Job enrichment lessons from AT&T, *Harvard Business Review,* 1973, 51, 96-106.

Gallagher, C. C., The human use of numbering systems, *Applied Ergonomics,* 1974, 5, 219-223.

Gentle, E. C., Jr., Keeping management up to the minute, *Computers and Data Processing,* May, 1964.

Gilb, T., *Reliable EDP Application Design,* Lund, Sweden: Studentlitteratur, 1973.

Gilb, T. and Weinberg, G. M., *Humanized Input: Techniques for Reliable Keyed Input,* Cambridge, Massachusetts: Winthrop Publishers, Inc., 1977.

Goldstein, I. L. and Mobley, W. H., Error and variability in the visual processing of dental radiographs, *Journal of Applied Psychology,* 1971, 55.

Gould, J. D., Visual factors in the design of computer-controlled CRT displays, *Human Factors,* August, 1968.

Hertzberg, F., One more time: How do you motivate employees? *Harvard Business Review,* 1968, 46, 53-62.

Hertzberg, F., Mausner, B. and Snyderman, B. B., *The Motivation to Work,* New York: John Wiley & Sons, Inc., 1967.

Hodge, M. H. and Field, M. M., Human coding processes, University of Georgia Report, January, 1970.

Holt, H. O., Beyond the system boundary, Presentation to the Advisory Committee in Charleston, South Carolina, April, 1969.

Hull, A. J., A letter-digit matrix of auditory confusions, *British Journal of Psychology,* 1963, 64, 579-585.

Jahnke, J. C. and Melton, A. W., Acoustic similarity and the Ranschburg Phenomenon, Proceedings of the 76th Annual American Psychological Association, 1968.

Kincaid, J. P., Fishburne, R. P., Rogers, R. L., and Chissom, B. S., Derivation of new readability formulas (Automated Readability Index, Fog Count, and Flesch Reading Ease Formula) for Navy enlisted personnel. Naval Training Command Research Branch Report 8-75. February 1975.

Kinkead, R. D. and Gonzalez, B. K., Human factors design recommendations for touch-operated keyboards, Minneapolis: Honeywell Inc., March, 1969.

Klemmer, E. T., Numerical error checking, *Journal of Applied Psychology,* 1959, 43.

Klemmer, E. T. and Lockhead, G. R., Productivity and errors in two keying tasks - A field study, *Journal of Applied Psychology,* 1962, 46, 401-408.

MacArthur, B. N., Accuracy of source data: Human error in hand transcription, AD623157, May, 1965.

Manley, C. W., Research and methodological considerations for human coding behavior, Presented at American Psychological Association Annual Meeting, 1973.

Martin, L. T., The input and correction interface in the Directory Project (DIR/ECT), Paper presented at 8th International Symposium on Human Factors in Telecommunications, Cambridge, England, September, 1977.

Mason, J. O., Jr. and Connelly, W. E., The application and reliability of the self-checking digit technique, *Management Advisor,* September-October, 1971.

Minor, F. J. and Revesman, S. L., Evaluation of input devices for a data setting task, *Journal of Applied Psychology,* 1962, 46, 332-336.

Minton, G., Inspection and correction error in data processing, *Journal of American Statistical Association,* 1969, 1256-1275.

Mitchell, V., Human operator performance in computer-based message switching systems: A case study, Bell Laboratories Report, November, 1979.

Moser, H. and Fotheringham, W., Number telling, AD260457, December, 1960.

Murdock, B. B. and Von Saal, W., Transpositions in short-term memory, *Journal of Experimental Psychology,* 1967, 74, 137-143.

Neisser, U. and Weene, P., A note on human recognition of hand-printed characters, *Information and Control,* 1960, 3.

Owsowitz, S. and Sweetland, A., Factors affecting coding errors, The Rand Corporation Report, Santa Monica, April, 1965.

Rabbitt, P. M. A., Three kinds of error-signalling responses in a serial-choice task, *Quarterly Journal of Experimental Psychology,* 1968.

Scales, E. M. and Chapanis, A., The effect on performance of tilting the toll-operator's keyset, *Journal of Applied Psychology,* 1954, 38, 452-456.

Smith, E. A., and Kincaid, P., Derivation and validation of the automated readability index for use with technical materials. *Human Factors,* 1970, 12, 457-464.

Smith, W. A., Accuracy of manual entries in data-collection devices, *Journal of Applied Psychology,* 1967, 54.

Stevenson, F. L., Beyond the system boundaries, Presented at the Lisle Engineering Conference, November, 1970.

Strub, M. H., Automated aids to on-line tactical data inputting, February, 1975.

Tirrell, J. A. and Klemmer, E. T., The effects of coding on keying rate, IBM Research Center Report, 1962.

Tomeski, E. A., Job enrichment and the computer: A neglected subject, *Computers and People,* November, 1974.

Van Cott, H. P. and Kinkade, R. G., Ed., *Human Engineering Guide to Equipment Design,* Washington, D.C.: U.S. Government Printing Office, 1972.

Vartabedian, A. G., Developing a graphic set for cathode ray tube display using a 7x9 dot pattern, *Applied Ergonomics,* March, 1973.

Wickelgren, W. A., Associative intrusions in short-term recall, *Journal of Experimental Psychology,* 1966, 72, 853-858.

INDEX

T

V

W